Elementary Calculus from an Advanced Standpoint

by

Dennis Morris

© Dennis Morris

Published by: Abane & Right

56 Coach Road

Brotton

Saltburn

TS12 2RP

United Kingdom

01287 678918

dennis355@btinternet.com

July 2016

Contents

Contents

Introduction

Calculus within the real numbers is highly developed and well understood. We are quite capable of handling both differentiation and integration within both the real numbers, $\mathbb{R}$, and within spaces comprised of many copies of the real numbers, $\mathbb{R}^n$. The expressions can be complicated and error prone, but we know what we are doing.

Our 4-dimensional space-time is a space of four copies of the real numbers, $\mathbb{R}^4$, fabricated together with six angles, and so we are correct to do calculus within our 4-dimensional space-time as if we were doing calculus in four copies of the real numbers. We have to allow that a space fabricated together from four separate copies of a division algebra, as opposed to being a single division algebra space like quaternion space, $\mathbb{H}$, has no algebraic structure to keep it 'flat' and that it might have 'curvature' which varies from point to point. We can take account of this curvature using the covariant derivative, which we also understand.

Great! we believe we are very capable people until we meet the division algebra spaces (also called spinor spaces). A division algebra space, like the well-known complex numbers, $\mathbb{C}$, or a quaternion space, $\mathbb{H}$, is not a space fabricated together as several copies of the real numbers. We cannot do calculus within a division algebra space in the same way that we do calculus in our 4-dimensional space-time.

Although mathematicians have been differentiating and integrating within the division algebra which we call the complex numbers, $\mathbb{C}$, for almost 200 years, the kind of differentiation, which we might call Cauchy-Riemann differentiation, is in some ways unsatisfactory, and, to be quite frank, in your author's opinion, it is not the most efficient way to do calculus within the complex numbers. Similarly, integration within the complex numbers, which we might call contour integration or Cauchy-Riemann integration, is in some ways unsatisfactory, and, to be quite frank, in your author's opinion, it is not the most efficient way to do calculus within the complex numbers. Perhaps your author is too strongly opinionated.

In this book, we present an alternative calculus within the complex numbers, $\mathbb{C}$. With this alternative, we can say that, at the very least, there is a different type of differentiation and integration within the complex numbers which is both more useful, and more satisfactory, and much simpler, than the Cauchy-Riemann calculus conventionally used within the complex numbers.

In this book, we also present the non-commutative calculus of the quaternions. The non-commutative calculus of the quaternions is, except for the non-commutativity, of the same ilk as the alternative calculus we have within the complex numbers.

Although most are unfamiliar to mathematicians, there are an infinite number of division algebras. They exist in every dimension. The complex numbers and the quaternions are just two better known examples of the infinite number of 'types of complex number', which are what we call division algebras. We really need a calculus, a way of differentiating and integrating, which is of the same nature in all division algebras. The alternative calculus presented in this book applies over all division algebras, including the non-commutative division algebras.

'The Fundamental Theorem of Calculus' states that integration is just the reverse (inverse, if you prefer) of differentiation. This 'Fundamental Theorem of Calculus' holds true over all division algebras. We integrate within the real numbers by reversing the differentiation, not always a simple task, and we will do the same within all other division algebras, an even harder task.

The reader is made aware that it is extremely audacious of your author to assert that mathematicians have been doing calculus inefficiently within the complex numbers for the last 200 years. The reader should be very cautious. Perhaps we should just accept that there is an alternative complex calculus and not upset folk. The reader will doubtless form her own opinion.

If the Cauchy-Riemann calculus is an inappropriate form of calculus, it is utterly remarkable that, for 200 years, mathematicians, including great mathematicians, blinded by their own over-confidence have been doing complex analysis inappropriately.

It will turn out that all the results of Cauchy-Riemann calculus are, almost by accident, correct even though the Cauchy-Riemann calculus is inappropriate.

The Cauchy-Riemann equations still have a use. They are needed for studying fluid flow and conformal mappings.

Chapter 1

A Quick Overview

A differential is a ratio of two variables; examples are:

$$\frac{dy(x)}{dx} \qquad \frac{df(x,y)+idg(x,y)}{dx+idy} \qquad (1.1)$$

A ratio is a product, and so a differential is a product.

Proper algebraic products exist within only division algebras[1], and so differentials exist within only division algebras.

Our 4-dimensional space-time is not a division algebra space. The differential in our 4-dimensional space-time is the sum of four 1-dimensional differentials each of which is within the 1-dimensional real numbers division algebra. This sum is presented as a 4-vector.

Because our 4-dimensional space-time is not a division algebra space, it has no algebraic structure to keep it flat[2]. The sum of four real differentials has to be adjusted to allow for our 4-dimensional space-time being curved. The adjusted sum is called the 4-dimensional space-time covariant derivative.

Our 4-dimensional space-time also holds three copies of each of the two division algebras which are the complex numbers (the Euclidean plane) and the hyperbolic complex numbers (the 2-dimensional space-time of special relativity). The differentials in these spaces are adjusted to allow for the orientation of the axes of these spaces varying locally over our 4-dimensional space-time. The adjusted differentials are manifest in our 4-dimensional space-time as the $U(1)$ gauge covariant differential and

---

[1] Division algebras are types of numbers like the real numbers or the complex numbers or the quaternions.
[2] All division algebra spaces are rigidly flat; they are so flat that they do not even have zero curvature.

as change of velocity between two points in space-time – acceleration of an inertial mass.

The two quaternion non-commutative differentials form a superimposed differential. This differential is adjusted to allow for the orientation of the axes of this space varying locally over our 4-dimensional space-time. The adjusted differential is the $SU(2)$ gauge covariant differential.

Chapter 2

Division Algebras

A division algebra is a type of numbers. Examples of division algebras are the 1-dimensional real numbers, $\mathbb{R}$, the 2-dimensional complex numbers, $\mathbb{C}$, and the 4-dimensional quaternions, $\mathbb{H}$. There are division algebras of any dimension, but most of these are not as well-known as the examples given above.

It is a fact that the elements of any division algebra, numbers that is, can be written as a set of square matrices. The size of the matrices are the dimension of the algebra. The real numbers are 1×1 matrices. Quaternions are 4×4 matrices.

The addition and multiplication operations of all commutative division algebras are no more than matrix addition and matrix multiplication; for example, the complex numbers, $\mathbb{C}$ are multiplied as:

$$(a+ib)(c+id) = ac - bd + i(bc + ad)$$

$$\begin{bmatrix} a & b \\ -b & a \end{bmatrix}\begin{bmatrix} c & d \\ -d & c \end{bmatrix} = \begin{bmatrix} ac-bd & bc+ad \\ -(bc+ad) & ac-bd \end{bmatrix} \tag{2.1}$$

For completeness, we present one of the two quaternion algebras in matrix form. This is the left-chiral quaternions:

$$a + \hat{i}b + jc + kd$$

$$\mathbb{H}_{Left-Chiral} = \begin{bmatrix} a & b & c & d \\ -b & a & -d & c \\ -c & d & a & -b \\ -d & -c & b & a \end{bmatrix} \tag{2.2}$$

There are also right-chiral quaternions. The use of both right-chiral and left-chiral quaternions is important within physics where is seems that a superimposition of both types of quaternions gives the electron field and the left-chiral neutrino field[3].

The addition operation of multiplicatively non-commutative division algebras is also matrix addition, but the multiplication operation of multiplicatively non-commutative division algebras, although based on matrix multiplication, is a little more complicated than simple matrix multiplication.

Summary:
Every type of number is a square matrix. Every type of multiplication is based on matrix multiplication – also known as linear multiplication.

---

[3] See : Dennis Morris : The Quaternion Dirac Equation

Chapter 3

A New Kind of Product

Within every division algebra, we have the two operations of multiplication and addition. Indeed, a division algebra is no more than a set of elements called numbers tied together by these two operations of addition and multiplication. Addition and multiplication are called the algebraic operations; these are the only algebraic operations; for example, taking the square root is not an algebraic operation. Technically, to be considered to be 'proper' algebraic operations, these two operations have to satisfy a number of axioms (have a number of properties). These axioms are things like multiplicative/additive associativity, multiplicative/additive closure, the existence of multiplicative inverses, the absence of zero-divisors, and multiplicative distributivity over addition. Note that multiplicative commutativity is not a required property.

A different approach is simply to realise that the elements (numbers) of all division algebras can be written as a set of matrices; for example the real numbers are 1×1 matrices and the quaternions are 4×4 matrices. It seems natural that these matrices should be tied together by matrix addition and matrix multiplication. Matrix addition and matrix multiplication have all the required properties necessary to be 'proper' algebraic operations.

Given the restrictions of matrix addition, we might think there can be only one type of algebraic addition – matrix addition; this is seemingly correct.

Given the restrictions of matrix multiplication, we might think there can be only one type of algebraic multiplication operation – matrix multiplication; this might not be correct. It seems that, at least in the case of non-commutative division algebras, there is a second kind of multiplication operation. This second kind of multiplication operation has matrix multiplication in its heart, but it is not straight-forward matrix multiplication. We will consider this alternative second multiplication operation.

A second type of multiplication:
How can it be that, during thousands of years of doing arithmetic, humankind has not previously discovered the nature of non-commutative multiplication? It is simply that until very recently no-one did non-commutative arithmetic. It is only the advent of particle physics over the last century that has brought non-commutative arithmetic into the minds of humankind.

The need for such consideration of an alternative second multiplication operation is evident within a non-commutative division algebra such as the quaternions. Within a non-commutative division algebra, the 'normal' matrix multiplication gives two products of any two elements of the algebra rather than just one product; we have:

$$Q_A Q_B = X \qquad Q_B Q_A = Y$$
$$X \neq Y$$

(3.1)

The order of the elements whose product we seek determines which of the two products we get. It seems unsatisfactory that we have a multiplication operation which produces two equally valid products rather than a single product and that we just 'throw away' one of those products. Since we have two products of every pair of elements within a non-commutative division algebra, surely we ought to use both these products rather than randomly discard one of them.

The traditional multiplication operation:
We traditionally multiply, which we signify by $\times$, two real numbers together as:

$$a \times b = ab$$
$$2 \times 3 = 6$$

(3.2)

How do we know this is the proper multiplication operation within the division algebra that is real numbers? Over thousands of years, we have become used to this form of multiplication; it is tried and well-tested; it works in all practical situations. However, numbers are in some sense outside of practical things. It seems that numbers exist in their own right over and above the practicalities of the material universe.

Suppose that, back in the dim distant past, there had been two 'first mathematicians' and that these two 'first mathematicians' had each defined multiplication of the real numbers in a different way and that both of these definitions of multiplication work in all practical situations and satisfy all the requirements necessary to be an algebraic operation. Then we would have two forms of multiplication operation each of which could equally claim to be tried and well-tested and a *bone-fide* multiplication operation. We would be unable to decide by practical means, or by reference to axioms, which of the two forms of multiplication was the correct multiplication operation within the real numbers.

Is there such an alternative second multiplication operation? Yes, it seems so. It seems, though, that there were not two 'first mathematicians'.

A weird multiplication operation:
We are guided by the success of non-commutative differentiation within physics; it is this that leads us to the alternative second multiplication operation; non-commutative differentiation is considered later in this book. That alternative second multiplication operation is:

$$a \times b = \begin{cases} \dfrac{1}{2}(ab + ba) \\ \dfrac{1}{2}(ab - ba) \end{cases} = \begin{Bmatrix} E_{\text{Prod}} \\ B_{\text{Prod}} \end{Bmatrix} \tag{3.3}$$

We also have:

$$b \times a = \left\{ \begin{array}{l} \dfrac{1}{2}(ba+ab) \\[2mm] \dfrac{1}{2}(ba-ab) \end{array} \right\} = \left\{ \begin{array}{l} E_{\text{Prod}} \\ -B_{\text{Prod}} \end{array} \right\} \qquad (3.4)$$

Notice that the order of the elements being multiplied together is important within this multiplication operation. We could have defined this operation to reverse the order, but that is mere semantics.

This multiplication operation produces two products, but, within a commutative algebra such as the real numbers or the complex numbers, the second of these, B_{Prod}, is always zero while the first of these, E_{Prod}, is always equal to what for millennia humankind has taken to be the product of commutative numbers.

Which of the two products, (3.2) or (3.3), is the correct algebraic product? Within a commutative division algebra, the question cannot be resolved and so it is of no consequence. However, there is a difference between these two multiplication operations within non-commutative division algebras.

Looking at (3.3), we see that both parts of the product include the two possible 'normal' products of two elements, ab & ba, of a non-commutative division algebra.

Interestingly, the B_{Prod} part of the product, (3.3), is chiral – it reverses sign when the order of the two elements of the division algebra are reversed. The E_{Prod} part of the product, (3.3), is not chiral.

Associativity of the alternative second product:
Let us examine this second multiplicative product in detail. We immediately discover one surprise. Within a non-commutative division algebra, for example the quaternions, the alternative second multiplication operation (3.3) is not associative.

We have:

$$a(bc) = \left\{ \begin{array}{l} a\dfrac{1}{2}(bc+cb) \\ a\dfrac{1}{2}(bc-cb) \end{array} \right\} = \left\{ \begin{array}{l} \dfrac{1}{4}(abc+acb+bca+cba) \\ \dfrac{1}{4}(abc-acb-bca+cba) \end{array} \right\} = \left\{ \begin{array}{l} E_{a(bc)} \\ B_{a(bc)} \end{array} \right\}$$

$$(ab)c = \left\{ \begin{array}{l} \dfrac{1}{2}(ab+ba)c \\ \dfrac{1}{2}(ab-ba)c \end{array} \right\} = \left\{ \begin{array}{l} \dfrac{1}{4}(abc+bac+cab+cba) \\ \dfrac{1}{4}(abc-bac-cab+cba) \end{array} \right\} = \left\{ \begin{array}{l} E_{(ab)c} \\ B_{(ab)c} \end{array} \right\}$$

$$(3.5)$$

We also have:

$$a(cb) = \left\{ \begin{array}{l} a\dfrac{1}{2}(cb+bc) \\ a\dfrac{1}{2}(cb-bc) \end{array} \right\} = \left\{ \begin{array}{l} \dfrac{1}{4}(acb+abc+cba+bca) \\ \dfrac{1}{4}(acb-abc-cba+bca) \end{array} \right\} = \left\{ \begin{array}{l} E_{a(cb)} = E_{a(bc)} \\ B_{a(cb)} = -B_{a(bc)} \end{array} \right\}$$

$$(ac)b = \left\{ \begin{array}{l} \dfrac{1}{2}(ac+ca)b \\ \dfrac{1}{2}(ac-ca)b \end{array} \right\} = \left\{ \begin{array}{l} \dfrac{1}{4}(acb+cab+bac+bca) \\ \dfrac{1}{4}(acb-cab-bac+bca) \end{array} \right\} = \left\{ \begin{array}{l} E_{(ac)b} \\ B_{(ac)b} \end{array} \right\}$$

$$(3.6)$$

And:

$$b(ac) = \left\{ \begin{array}{l} b\dfrac{1}{2}(ac+ca) \\ b\dfrac{1}{2}(ac-ca) \end{array} \right\} = \left\{ \begin{array}{l} \dfrac{1}{4}(bac+bca+acb+cab) \\ \dfrac{1}{4}(bac-bca-acb+cab) \end{array} \right\} = \left\{ \begin{array}{l} E_{b(ac)} = E_{(ac)b} \\ B_{b(ac)} = -B_{(ac)b} \end{array} \right\}$$

$$(ba)c = \left\{ \begin{array}{l} \dfrac{1}{2}(ba+ab)c \\ \dfrac{1}{2}(ba-ab)c \end{array} \right\} = \left\{ \begin{array}{l} \dfrac{1}{4}(bac+abc+cba+cab) \\ \dfrac{1}{4}(bac-abc-cba+cab) \end{array} \right\} = \left\{ \begin{array}{l} E_{(ba)c} = E_{(ab)c} \\ B_{(ba)c} = -B_{(ab)c} \end{array} \right\}$$

$$(3.7)$$

And:

$$b(ca) = \begin{cases} b\dfrac{1}{2}(ca+ac) \\[2ex] b\dfrac{1}{2}(ca-ac) \end{cases} = \begin{cases} \dfrac{1}{4}(bca+bac+cab+acb) \\[2ex] \dfrac{1}{4}(bca-bac-cab+acb) \end{cases} = \begin{cases} E_{b(ca)} = E_{(ac)b} \\[2ex] B_{b(ca)} = B_{(ac)b} \end{cases}$$

$$(bc)a = \begin{cases} \dfrac{1}{2}(bc+cb)a \\[2ex] \dfrac{1}{2}(bc-cb)a \end{cases} = \begin{cases} \dfrac{1}{4}(bca+cba+abc+acb) \\[2ex] \dfrac{1}{4}(bca-cba-abc+acb) \end{cases} = \begin{cases} E_{(bc)a} = E_{a(bc)} \\[2ex] B_{(bc)a} = -B_{a(bc)} \end{cases}$$

(3.8)

And:

$$c(ab) = \begin{cases} c\dfrac{1}{2}(ab+ba) \\[2ex] c\dfrac{1}{2}(ab-ba) \end{cases} = \begin{cases} \dfrac{1}{4}(cab+cba+abc+bac) \\[2ex] \dfrac{1}{4}(cab-cba-abc+bac) \end{cases} = \begin{cases} E_{c(ab)} = E_{(ab)c} \\[2ex] B_{c(ab)} = -B_{(ab)c} \end{cases}$$

$$(ca)b = \begin{cases} \dfrac{1}{2}(ca+ac)b \\[2ex] \dfrac{1}{2}(ca-ac)b \end{cases} = \begin{cases} \dfrac{1}{4}(cab+acb+bca+bac) \\[2ex] \dfrac{1}{4}(cab-acb-bca+bac) \end{cases} = \begin{cases} E_{(ca)b} = E_{(ac)b} \\[2ex] B_{(ca)b} = -B_{(ac)b} \end{cases}$$

(3.9)

And:

$$c(ba) = \begin{cases} c\dfrac{1}{2}(ba+ab) \\[2ex] c\dfrac{1}{2}(ba-ab) \end{cases} = \begin{cases} \dfrac{1}{4}(cba+cab+bac+abc) \\[2ex] \dfrac{1}{4}(cba-cab-bac+abc) \end{cases} = \begin{cases} E_{c(ba)} = E_{(ab)c} \\[2ex] B_{c(ba)} = B_{(ab)c} \end{cases}$$

$$(cb)a = \begin{cases} \dfrac{1}{2}(cb+bc)a \\[2ex] \dfrac{1}{2}(cb-bc)a \end{cases} = \begin{cases} \dfrac{1}{4}(cba+bca+acb+abc) \\[2ex] \dfrac{1}{4}(cba-bca-acb+abc) \end{cases} = \begin{cases} E_{(cb)a} = E_{a(bc)} \\[2ex] B_{(cb)a} = B_{a(bc)} \end{cases}$$

(3.10)

We see that, up to sign, we have six possible products of three quaternions.

$$E_{a(bc)}, \qquad E_{(ab)c}, \qquad E_{(ac)b}$$
$$\pm B_{a(bc)}, \qquad \pm B_{(ab)c}, \qquad \pm B_{(ac)b}$$

$$(3.11)$$

Of course, there are six possible normal matrix products of three quaternions using the normal matrix product corresponding to:

$$abc, \quad acb, \quad bac, \quad bca, \quad cab, \quad cba \qquad (3.12)$$

None of these are equal except that the real variables, $X_{[1,1]}$,[4] are equal if the order of the products are cyclically the same:

$$abc_{[1,1]} = bca_{[1,1]} = cab_{[1,1]}$$
$$acb_{[1,1]} = bac_{[1,1]} = cba_{[1,1]}$$

$$(3.13)$$

We see that all six of these possible normal matrix products occur in the two products within (3.5). If we were to insist upon using only one of the two products in (3.5), we would lose two of the possible normal matrix products of three quaternions.

Notes on the side:
We cannot make this second product seemingly associative by imposing an 'order' of doing multiplication. We cannot, say, agree that when presented with three non-commutative numbers as in (3.5), then we always evaluate the product from right to left (it could equally well be from left to right) so that a product of more than two non-commutative numbers would always be $a(bc)$ and never $(ab)c$ for to do this is to lose two of the possible normal matrix products of three quaternions. If we are to include all six possible normal matrix products of three quaternions, then we must use both of the products in (3.5).[5]

We see that, since the real numbers are commutative, the two calculations above, (3.5), are equal within the real numbers, and so this

---

[4] The real variables appear on the leading diagonals of the matrix notation; hence the subscript.
[5] We wonder if this is anything to do with the need to mandate the order in which operators act within QED.

new E/B product is associative within the real numbers. The same is true for any kind of commutative division algebra in which the variables $\{a,b,c\}$ are elements of that algebra.

However, if we were dealing with non-commutative entities like the quaternions, the E/B product would not be associative since the two calculations above, (3.5), would be unequal. It is quite surprising that, when using the E/B product, we lose associativity if we lose commutativity.

Reconsidering associativity:
At first thought, this lack of associativity might deter the reader from using the E/B product. If we have lost associativity, we have surely lost an essential part of the structure necessary to form a division algebra; please read on.

We think it entirely sensible that a non-commutative division algebra like the quaternions should have two products, AB & BA. Keeping order, lack of associativity does only the same, $(AB)C$ & $A(BC)$. Perhaps we hold associativity in too high esteem if lacking it does no more than is done by non-commutativity.

The associator:
Mathematicians have previously considered a kind of non-associative product called the associator; we briefly mention this.

All division algebras are associative by definition. We will need to write of 'aspirant algebras'; these are non-associative but otherwise would be considered to be division algebras. So, crudely, an 'aspirant algebra' is a non-associative division algebra. There is no such term as 'aspirant algebra' outside of this book.

The associator[6] of an 'aspirant algebra' is defined as:

$$A(X,Y,Z) = (XY)Z - X(YZ) \qquad (3.14)$$

Looking at (3.5) which is $a(bc) \& (ab)c$, we have the associators as:

$$(ab)c - a(bc) = \begin{cases} bac + cab - acb - bca = -B_{(ac)b} \\ -(bac + cab - acb - bca) = B_{(ac)b} \end{cases} \qquad (3.15)$$

These are just a sign different; we care not for the $\frac{1}{4}$. We see how remarkably entwined are these products.

We will not consider the associator further in this book.

Summary of associativity:
We will be using the alternative second product. We need the alternative second product to be able to form the non-commutative differential which inspired it.

We take the view that lack of associativity does no more than lack of commutativity in that it simply increases the number of possible products. We can handle this easily. We take the view that the alternative second product is tidier than the normal product in that we capture every order of the non-commutative elements within it.

The chirality of the alternative second product is particularly interesting; it eventually leads to neutrinos being left-handed.

Distributivity:
Distributivity of multiplication over addition is:

---

[6] See : J. P. Ward : Quaternions and Cayley Numbers page 178 quoting Curtis, C : The Four and Eight Square Problem and Division Algebras : Studies in Modern Algebra : Prentice-Hall Inc (1963)

$$a(b+c) = ab + ac \tag{3.16}$$

Using the alternative second product, we have:

$$a(b+c) = \left\{ \begin{array}{l} \left|\dfrac{1}{2}\big(a(b+c)+(b+c)a\big)\right| \\ \left|\dfrac{1}{2}\big(a(b+c)-(b+c)a\big)\right| \end{array} \right\} = \left\{ \begin{array}{l} \left|\dfrac{1}{2}(ab+ac+ba+ca)\right| \\ \left|\dfrac{1}{2}(ab+ac-ba-ca)\right| \end{array} \right\}$$

$$ab + ac = \left\{ \begin{array}{l} \left|\dfrac{1}{2}(ab+ba)\right| \\ \left|\dfrac{1}{2}(ab-ba)\right| \end{array} \right\} + \left\{ \begin{array}{l} \left|\dfrac{1}{2}(ac+ca)\right| \\ \left|\dfrac{1}{2}(ac+ca)\right| \end{array} \right\} = \left\{ \begin{array}{l} \left|\dfrac{1}{2}(ab+ac+ba+ca)\right| \\ \left|\dfrac{1}{2}(ab+ac-ba-ca)\right| \end{array} \right\}$$

$$\tag{3.17}$$

The alternative second product is distributive over addition.

Multiplicative identity:

$$1a = \left\{ \begin{array}{l} \left|\dfrac{1}{2}(1a+a1)\right| \\ \left|\dfrac{1}{2}(1a-a1)\right| \end{array} \right\} = \left\{ \begin{array}{l} a \\ 0 \end{array} \right\} \tag{3.18}$$

We have 1 as the multiplicative identity of the E-product but not of the B-product. This failure of the B-product is just the zero nature of the B-product for commutative variables. The multiplicative identity is a real number, and so it commutes with every other element of the algebra.

Note: It is possible to consider the multiplicative identity to be the unit 'sphere' in which case the multiplicative identity is the rotation matrix of the algebra – this would be the circle in the complex plane, $\mathbb{C}$. We will not go down that route.

Multiplicative closure:
Since the E-product and the B-product are just sums of the traditional product, any set of objects which are multiplicatively closed under the

traditional product will be multiplicatively closed under the E/B product. This is all we need for multiplicative closure. In short, the E-product and the B-product are the same kind of objects (quaternions for example) as the two objects which are combined together in the E/B product.

Conjugation:

All division algebras have a polar form which is a real number (radial variable) multiplied by a rotation matrix. Since real numbers commute with everything, we have:

$$[radial][Rot] = [Rot][Radial]$$

$$\begin{bmatrix} r & 0 \\ 0 & r \end{bmatrix}\begin{bmatrix} \cos\theta & \sin\theta \\ -\sin\theta & \cos\theta \end{bmatrix} = \begin{bmatrix} \cos\theta & \sin\theta \\ -\sin\theta & \cos\theta \end{bmatrix}\begin{bmatrix} r & 0 \\ 0 & r \end{bmatrix} \quad (3.19)$$

We have chosen the commutative complex numbers, $\mathbb{C}$, to illustrate this point for presentational ease, but the point is true for non-commutative division algebras.

Conjugation is just reverse rotation; for example the conjugate of:

$$\begin{bmatrix} r & 0 \\ 0 & r \end{bmatrix}\begin{bmatrix} \cos\theta & \sin\theta \\ -\sin\theta & \cos\theta \end{bmatrix} \quad (3.20)$$

is (note that θ is the imaginary variable):

$$\begin{bmatrix} r & 0 \\ 0 & r \end{bmatrix}\begin{bmatrix} \cos(-\theta) & \sin(-\theta) \\ -\sin(-\theta) & \cos(-\theta) \end{bmatrix} = \begin{bmatrix} r & 0 \\ 0 & r \end{bmatrix}\begin{bmatrix} \cos\theta & -\sin\theta \\ \sin\theta & \cos\theta \end{bmatrix} \quad (3.21)$$

Within both the complex numbers and the quaternion algebras, this reverse rotation can be expressed in the Cartesian form of the algebra as a simple change of sign of the imaginary variables, but these algebras are exceptional in this regard. In most division algebras, the conjugate Cartesian form is more complicated than a simple sign change. In such algebras, the easiest way to conjugate an element of the algebra is to reverse the sign of the angle variables[7] in the rotation matrix (polar form).

It is a property of all division algebras, commutative and non-commutative, that the product of a rotation matrix and the reverse of that rotation matrix is the identity:

$$\begin{bmatrix} \cos\theta & \sin\theta \\ -\sin\theta & \cos\theta \end{bmatrix} \begin{bmatrix} \cos\theta & -\sin\theta \\ \sin\theta & \cos\theta \end{bmatrix} = \begin{bmatrix} 1 & 0 \\ 0 & 1 \end{bmatrix} \qquad (3.22)$$

Intuitively, this seems obvious, but intuition can mislead us. Mathematically, this is connected to the fact that the exponential of a matrix with zero trace is a matrix with determinant unity and that all rotation matrices, being the exponentials of matrices with zeros on the leading diagonal, have determinant unity. It is expressed as the determinant of all rotation matrices being unity which we call a trigonometric identity similar to:

$$\cos^2\theta + \sin^2\theta = 1$$
$$v_A^3(\theta,\phi) + v_B^3(\theta,\phi) + v_C^3(\theta,\phi) - 3v_A(\theta,\phi)v_B(\theta,\phi)v_C(\theta,\phi) = 1$$
$$(3.23)$$

Wherein we have used both the 2-dimensional complex numbers, $\mathbb{C}$, and a 3-dimensional division algebra[8] to illustrate this fact.

The important point is that in any division algebra, every element of that algebra commutes with its conjugate:

$$QQ^* = Q^*Q = r^n \qquad (3.24)$$

This is a statement that the space of the division algebra, the spinor space, is rotationally symmetric. The power of the radial variable is the dimension of the algebra.

---

[7] Higher dimensional division algebras have more than one angle variable in their polar forms.
[8] See: Dennis Morris : Complex Numbers The Higher Dimensional Forms. The higher dimensional trigonometric functions have an 'oddness' and an 'evenness' appropriate to the dimension of their space; these reflective symmetries are almost completely unexplored mathematics.

Inclusion of multiplicative inverses:

The multiplicative inverse of any element is:

$$\frac{1}{a} = \left\{ \begin{array}{c} \frac{1}{2}\left(1\frac{1}{a}+\frac{1}{a}1\right) \\ \frac{1}{2}\left(1\frac{1}{a}-\frac{1}{a}1\right) \end{array} \right\} = \left\{ \begin{array}{c} \frac{1}{2}\left(\frac{a*}{aa*}+\frac{a*}{aa*}\right) \\ 0 \end{array} \right\} = \left\{ \begin{array}{c} \frac{a*}{r^n} \\ 0 \end{array} \right\} \tag{3.25}$$

We see that the multiplicative inverse of the E-product is just the conjugate divided by the radial distance and the there is no multiplicative inverse of the B-product.

Squares:

Let us calculate a square of an element of the division algebra using the alternative second product. We take the square of the element to be the product of the elements and its conjugate:

$$a^2 = \left\{ \begin{array}{c} \frac{1}{2}(a*a+aa*) \\ \frac{1}{2}(a*a-aa*) \end{array} \right\} = \left\{ \begin{array}{c} a^2 \\ 0 \end{array} \right\} \tag{3.26}$$

We see that the B-product gives a square that is zero. We can formulate QED with path integrals. This works correctly for bosons using conventional mathematics, but, to deal with fermions using path integrals, we have to use Grassman variables.

Grassman variables are often called Grassman numbers, but they are not really numbers because they do not have the properties necessary to be a division algebra. Even so, we can think of Grassman variables, $\{\eta,\xi\}$, as objects such that:

$$\eta\xi = -\xi\eta \tag{3.27}$$

This is taken to imply that:

$$\eta^2 = 0 \tag{3.28}$$

Grassman variables have the property that they square to zero; this is a good reason why they are not variables within a division algebra. Conventionally, this property is imposed upon the Grassman variables with a twelve-pound hammer; it does not arise naturally within Grassman algebra. It does arise naturally within our alternative second product.

Zero divisors:

One of the division algebra axioms is the absence of zero divisors. The mantra goes that a set of mathematical objects cannot be a division algebra if the product of two non-zero objects is zero. In symbols, this is:

$$a \times b \neq 0 \quad unless \quad a = 0 \ or \ b = 0 \ or \ a = b = 0 \qquad (3.29)$$

This axiom is really a statement of the corresponding observed property of the real numbers. This axiom assumes the traditional form of product. If we are to adopt the E-product and B-product form of product, then the absence of zero divisors axiom needs to be modified to account for the different form of product.

With the E/B product, the absence of zero divisors axiom becomes 'we cannot have both E-product and B-product equal to zero unless at least one of the objects of the algebra is zero':

$$Both \ \left\{ \begin{matrix} \frac{1}{2}(ab+ba) \\ \frac{1}{2}(ab-ba) \end{matrix} \right\} \neq 0 \quad unless \quad a = 0 \ or \ b = 0 \ or \ a = b = 0$$

$$(3.30)$$

A zero E-product implies that $ba = -ab$; this implies the B-product is ab which cannot be zero unless $a = 0$ or $b = 0$ or $a = b = 0$.

It is quite easy to find quaternions which have a zero B-product. The square of a quaternion (the quaternion multiplied by its conjugate) is a quaternion with a zero B-product but it has non-zero E-product. An example of two quaternions with zero E-product is the quaternions, in non-matrix notation:

$$(0+0i+j+0k)(0+i+0j+0k) \qquad (3.31)$$

These have E-product and B-product:

$$E_{\text{Prod}} = \frac{1}{2}(ji+ij) = \frac{1}{2}(-k+k) = 0$$

$$B_{\text{Prod}} = \frac{1}{2}(ji-ij) = \frac{1}{2}(-k-k) = -k \qquad (3.32)$$

Summary:

We have found an alternative second product of two elements of a division algebra. This second product has all the attributes we require of multiplication albeit in an unconventional way.

This second product incorporates the matrix product as terms within the second product.

For a commutative division algebra, the second product gives the same results as the conventional product which is simple matrix multiplication.

Within a non-commutative division algebra, we lose associativity if we use the alternative second product but this does no more than produce more products. The number of extra products is the same as the number of conventional matrix multiplication products, and so we have lost nothing by choosing the alternative second product.

This second product has interesting properties like the chirality of the B-product and the 'square to zero' nature of the B-product. These might be connected to QED.

We know that non-commutative differentiation produces the correct results within physics. This second product we have discovered is exactly what we need to use non-commutative differentiation.

Furthermore, we need a non-commutative product to do calculus within a non-commutative division algebra. All differentiation is a product:

$$dy \frac{1}{dx} \qquad (3.33)$$

Non-commutative differentiation must be a non-commutative product. We need a non-commutative product from somewhere.

A question for the reader:

Now, can you be sure that the normal matrix product is the 'proper' multiplication operation within a division algebra or do you have to allow that the E/B product might be the 'proper' multiplication operation within a division algebra? Consideration of the practicalities associated with physics seems to favour the E/B product at least within non-commutative division algebras.

Chapter 4

Commutative Differentiation

We begin our look at differentiation by going back to the basics.

Differentiation from first principles:
Initially, we work within the real numbers. We will differentiate a simple function from first principles. This is standard stuff taught at secondary schools. We have that simple function:

$$y = x^2 \tag{4.1}$$

We let the y variable increase by a small amount and the x variable increases by a commensurate small amount:

$$\begin{aligned} y + \delta y &= (x + \delta x)^2 \\ &= x^2 + 2x\delta x + \delta x^2 \end{aligned} \tag{4.2}$$

We wave our arms around to justify discarding the δx^2 as being insignificant, and we get:

$$\delta y \frac{1}{\delta x} = 2x \tag{4.3}$$

Now, on the LHS of (4.3), we have a product. Within the real numbers, using the real number product with the two real variables, we have:

$$\frac{\delta y}{\delta x} = 2x \tag{4.4}$$

Waving our arms around in a desire to distract the audience from our lack of understanding of infinitesimals, we say something like 'as $\delta y \to 0$, so $\dfrac{\delta y}{\delta x} \to \dfrac{dy}{dx}$', and we have our differential:

$$\frac{dy}{dx} = 2x \tag{4.5}$$

The differential is the product of a variable and the inverse of a variable.

Great! Marvellous! We've all done this before, but now let us look at the above from an advanced standpoint.

Differentiation within a division algebra:
A differential is a ratio – how one variable changes with respect to another variable. To differentiate, we need to form a product similar to $\delta y \frac{1}{\delta x}$. Products exist in only algebraic structures that have a multiplication operation, and so we can differentiate within only division algebras.

Note: There are some algebraic structures, like Clifford algebras, which, when formulated properly[9], are really division algebras in disguise. There are some structures, like vectors in our 4-dimensional space-time, which are not division algebras but are assemblages of division algebras.

Different division algebras have different differentiation:
Different division algebras have different types of variables; for example, a complex variable is different from a real variable and different also from a quaternion variable. The differential is a product of a variable and the inverse of a variable (the inverse of a variable is always a variable within a division algebra), and so differentiation within a given division algebra is a product of two variables of that given division algebra; for example, a complex number differential will be of the commutative form:

---

[9] See : Dennis Morris : The Naked Spinor

$$(a+ib)\left(\frac{1}{c+id}\right) = \left(\frac{1}{c+id}\right)(a+ib) \qquad (4.6)$$

Different division algebras have different multiplication operations. For example, the quaternions, $\mathbb{H}$, have a non-commutative form of multiplication whilst the complex numbers, $\mathbb{C}$, have a commutative form of multiplication. Since the differential is a product like (4.3), the differential within a given division algebra will be of the same form as the multiplication operation of that division algebra. This means we will have a commutative form of differentiation within the complex numbers, $\mathbb{C}$, and a non-commutative form of differentiation within the quaternion algebras, $\mathbb{H}$. Conventionally, within a non-commutative algebra, such as the quaternions, we have at least two candidates for the differential:

$$\frac{1}{Q_A}Q_B \neq Q_B\frac{1}{Q_A} \qquad (4.7)$$

All multiplication is linear:

Every division algebra has a finite group at its heart. Every finite group of order n can be written as a set of $n \times n$ permutation matrices[10]. When each permutation matrix is multiplied by a different real variable and these permutation matrices are added together, they form the basic matrix form of the division algebras of the underlying group. In short, this means that the multiplication operation of every division algebra is based on matrix multiplication. The different division algebras within the finite group which underlies them differ in the distribution of minus signs within the permutation matrices[11].

Since the multiplication operation of every division algebra is based on matrix multiplication, that's linear multiplication to posh folk, every

---

[10] A permutation matrix is a square matrix with a single 1 in each row and a single 1 in each column.

[11] See : Dennis Morris : Complex Numbers The Higher Dimensional Forms – 2nd Edition.

differentiation has a matrix multiplication form; we will meet these soon.

Conventional differentiation within the complex numbers:
Conventional differentiation within the complex numbers, $\mathbb{C}$, is Cauchy-Riemann differentiation. Conventionally, differentiation of a complex variable, $z = x + iy$ is done by treating the complex variable, z, as if it were a real variable; for example:

$$f(z) = z^4 + 5z^3 + 3z^2 + 3z + 4$$
$$\frac{df}{dz} = 4z^3 + 15z^2 + 6z + 3 \tag{4.8}$$

Similarly, powers of a complex variable are treated as if they were powers of a real number; conventionally we take:

$$z = x + iy$$
$$z^2 = (x + iy)(x + iy) = x^2 - y^2 + 2ixy \tag{4.9}$$

Note: It might make more sense to take:

$$z^2 = (x + iy)(x - iy) = x^2 + y^2 \tag{4.10}$$

but this is not the convention.

Clearly, a function of a complex variable, $z = x + iy$, is a function of two variables, $\{x, y\}$. Consider the complex function:

$$f = z^3 = (x + iy)^3$$
$$= (x^3 - 3xy^2) + i(3x^2 y - y^3) \tag{4.11}$$
$$= u(x, y) + iv(x, y)$$

Since the derivative is a product and all division algebras are multiplicatively closed, the derivative has to be an element of the division algebra; in this case, the derivative of (4.11) has to be a complex number.

The conventional derivative of (4.11) is:

$$\frac{df}{dz} = 3z^2 = \left(3x^2 - 3y^2\right) + i\left(6xy\right) \tag{4.12}$$

In matrix form, these complex numbers, (4.11) & (4.12), are:

$$\mathbb{C} = \begin{bmatrix} a & b \\ -b & a \end{bmatrix} \tag{4.13}$$

$$f = \begin{bmatrix} x^3 - 3xy^2 & 3x^2y - y^3 \\ -\left(3x^2y - y^3\right) & x^3 - 3xy^2 \end{bmatrix} \tag{4.14}$$

$$\frac{df}{dz} = \begin{bmatrix} 3x^2 - 3y^2 & 6xy \\ -6xy & 3x^2 - 3y^2 \end{bmatrix} \tag{4.15}$$

We form the partial derivatives of (4.11) to get:

$$\frac{\partial u}{\partial x} = 3x^2 - 3y^2 \qquad \frac{\partial u}{\partial y} = -6xy$$

$$\frac{\partial v}{\partial x} = 6xy \qquad \frac{\partial v}{\partial y} = 3x^2 - 3y^2 \tag{4.16}$$

We have four derivatives, but a complex number is 2-dimensional; it has only two bits; it makes no sense to discard two of the four derivatives; what are we to do with the 'extra' derivatives?

Differentiating (forming a ratio of) a real function by a real variable gives a real function; we know that from the calculus of the real numbers. Similarly, the ratio of an imaginary function and an imaginary variable will be a real function – the i s cancel. Similarly, the derivative of a real function and an imaginary variable will be an imaginary function and the derivative of an imaginary function and a real variable will be an imaginary function. Thus, we know:

$$\frac{\partial u}{\partial x} \in \mathbb{R} \qquad\qquad \frac{\partial u}{\partial y} \in \text{Im}$$

$$\frac{\partial v}{\partial x} \in \text{Im} \qquad\qquad \frac{\partial v}{\partial y} \in \mathbb{R} \tag{4.17}$$

We can put these into a matrix to match (4.15):

$$\frac{df}{dz} = \begin{bmatrix} \dfrac{\partial u}{\partial x} & \dfrac{\partial v}{\partial x} \\[1.5em] \dfrac{\partial u}{\partial y} & \dfrac{\partial v}{\partial y} \end{bmatrix} \quad or \quad \frac{df}{dz} = \begin{bmatrix} \dfrac{\partial u}{\partial x} & \dfrac{\partial u}{\partial y} \\[1.5em] \dfrac{\partial v}{\partial x} & \dfrac{\partial v}{\partial y} \end{bmatrix}$$

$$\frac{df}{dz} = \begin{bmatrix} \dfrac{\partial v}{\partial y} & \dfrac{\partial v}{\partial x} \\[1.5em] \dfrac{\partial u}{\partial y} & \dfrac{\partial u}{\partial x} \end{bmatrix} \quad or \quad \frac{df}{dz} = \begin{bmatrix} \dfrac{\partial v}{\partial y} & \dfrac{\partial u}{\partial y} \\[1.5em] \dfrac{\partial v}{\partial x} & \dfrac{\partial u}{\partial x} \end{bmatrix} \tag{o(4.18)}$$

Clearly, looking at the form of (4.13), such a matrix as (4.18) is a complex number only if:

$$\frac{\partial u}{\partial x} = \frac{\partial v}{\partial y} \qquad \& \qquad \frac{\partial v}{\partial x} = -\frac{\partial u}{\partial y} \tag{4.19}$$

These are the Cauchy-Riemann equations. It is conventionally taken that a complex function is differentiable only if the Cauchy-Riemann equations are satisfied because only if the Cauchy Riemann equations are satisfied can the four differentials be cut down to two differentials and thereby form a complex number. Since the differential is a product, and we must have multiplicative closure within a division algebra, the Cauchy Riemann equations must be satisfied if the algebra is not to fall apart – very sensible.

Well! this is conventional differentiation within the complex numbers. We might call this Cauchy-Riemann differentiation or just conventional differentiation. It is a differentiation of some form, but your author opines that it is not 'proper' differentiation within the complex numbers.

The phrase 'differentiable only if the Cauchy-Riemann equations are satisfied' applies to only differentiation as defined conventionally such as shown in (4.8). Perhaps this conventional differentiation is not the only type of differentiation within the complex numbers. If we choose a different type of differentiation, then the Cauchy Riemann equations might not be a necessary condition.

Integration:
We point out that in conventional complex analysis, a function is conventionally integrable if and only if it is Cauchy-Riemann differentiable.

Differentiation within the complex numbers, $\mathbb{C}$ - a different view:
The reader might be tempted to do differentiation within the complex numbers division algebra, $\mathbb{C}$, from first principles in a manner very similar to how we differentiated within the real numbers from first principles above, (4.2).

Using common sense, we will expect get something like $\dfrac{d(a+ib)}{d(x+iy)}$

within which $a(x,y)$ & $b(x,y)$ are both functions of $\{x,y\}$. Let us evaluate this using the complex number multiplication:

$$\begin{aligned}
\frac{da+idb}{dx+idy} &= \frac{(da+idb)(dx-idy)}{dx^2+dy^2} \\[2mm]
&= \frac{dadx+dbdy+i(-dady+dbdx)}{dx^2+dy^2} \\[2mm]
&= \frac{da}{dx+\dfrac{dy^2}{dx}} + \frac{db}{\dfrac{dx^2}{dy}+dy} - i\frac{da}{\dfrac{dx^2}{dy}+dy} + i\frac{db}{dx+\dfrac{dy^2}{dx}}
\end{aligned}$$

(4.20)

Ignoring the $\{dx^2, dy^2\}$ terms, we get:

$$\frac{d(a+ib)}{d(x+iy)} = \frac{da}{dx} + \frac{db}{dy} + i\left(\frac{db}{dx} - \frac{da}{dy}\right) \tag{4.21}$$

We see that we have reduced the four differentials $\left\{\frac{\partial a}{\partial x}, \frac{\partial a}{\partial y}, \frac{\partial b}{\partial x}, \frac{\partial b}{\partial y}\right\}$

to a real bit and an imaginary bit.

In matrix form:

$$\frac{d(a+ib)}{d(x+iy)} = \begin{bmatrix} \dfrac{da}{dx} + \dfrac{db}{dy} & \dfrac{db}{dx} - \dfrac{da}{dy} \\[2ex] -\left(\dfrac{db}{dx} - \dfrac{da}{dy}\right) & \dfrac{da}{dx} + \dfrac{db}{dy} \end{bmatrix} \tag{4.22}$$

This is very different from (4.18) above. What we have done (4.20) to (4.22) is quite simply differentiation from first principles within the complex numbers, but it effectively undermines a whole well established area of mathematics based upon the conventional Cauchy-Riemann way of differentiating within the complex numbers. It is for such as this that mathematicians such as your author are burned at the stake or thrown from the roof of a university mathematics building. It's the simplicity of it that the establishment does not like.

Matrix differentiation:

A much easier way of differentiating the complex numbers is to use the matrix form of the complex numbers. Since the multiplication operation of every division algebra is based on matrix multiplication, this technique can be applied to every division algebra. For higher dimensional division algebras, this technique is cumbersome involving many large matrices, but it is simple, clear, and seemingly correct[12].

We will differentiate the complex function:

---

[12] Form your own opinion.

$$\begin{bmatrix} f(x,y) & g(x,y) \\ -g(x.y) & f(x,y) \end{bmatrix}$$

(4.23)

With respect to the complex variable:

$$\begin{bmatrix} x & y \\ -y & x \end{bmatrix}$$

(4.24)

We have:

$$\frac{\partial \begin{bmatrix} f(x,y) & g(x,y) \\ -g(x.y) & f(x,y) \end{bmatrix}}{\partial \begin{bmatrix} x & y \\ -y & x \end{bmatrix}}$$

(4.25)

$$= \frac{\partial \begin{bmatrix} f(x,y) & 0 \\ 0 & f(x,y) \end{bmatrix}}{\partial \begin{bmatrix} x & 0 \\ 0 & x \end{bmatrix}} + \frac{\begin{bmatrix} 0 & 1 \\ -1 & 0 \end{bmatrix} \partial \begin{bmatrix} g(x,y) & 0 \\ 0 & g(x,y) \end{bmatrix}}{\partial \begin{bmatrix} x & 0 \\ 0 & x \end{bmatrix}}$$

(4.26)

$$+ \frac{\partial \begin{bmatrix} f(x,y) & 0 \\ 0 & f(x,y) \end{bmatrix}}{\begin{bmatrix} 0 & 1 \\ -1 & 0 \end{bmatrix} \partial \begin{bmatrix} y & 0 \\ 0 & y \end{bmatrix}} + \frac{\begin{bmatrix} 0 & 1 \\ -1 & 0 \end{bmatrix} \partial \begin{bmatrix} g(x,y) & 0 \\ 0 & g(x,y) \end{bmatrix}}{\begin{bmatrix} 0 & 1 \\ -1 & 0 \end{bmatrix} \partial \begin{bmatrix} y & 0 \\ 0 & y \end{bmatrix}}$$

We have reduced everything to real number differentiation which we understand. Tidying this, (4.26), gives:

$$\frac{\partial \begin{bmatrix} f(x,y) & g(x,y) \\ -g(x.y) & f(x,y) \end{bmatrix}}{\partial \begin{bmatrix} x & y \\ -y & x \end{bmatrix}} = \begin{bmatrix} \dfrac{\partial f}{\partial x} + \dfrac{\partial g}{\partial y} & \dfrac{\partial g}{\partial x} - \dfrac{\partial f}{\partial y} \\ -\left(\dfrac{\partial g}{\partial x} - \dfrac{\partial f}{\partial y}\right) & \dfrac{\partial f}{\partial x} + \dfrac{\partial g}{\partial y} \end{bmatrix}$$

(4.27)

$$= \begin{bmatrix} Div & Curl \\ -Curl & Div \end{bmatrix}$$

This is the same as (4.22) above.

A complex potential:
We see that the differential of a complex function is a divergence and a curl, (4.27). What kind of thing is it that has a divergence and a curl? It is a potential like the potential that leads to an electric field and a magnetic field by differentiation.

So, if we have a complex function over two of our space-time dimensions, we have a potential over two of our space-time dimensions. Take three such complex functions set at right-angles to each other, and we have a 3-dimensional potential.

In due course, we will meet gauge covariant differentiation, and we will examine the $U(1)$ gauge covariant differential. This is the differential of a complex function over our 4-dimensional space-time. Keep in mind that a complex function is a potential.

Looking at (4.27), we notice that the differential is of the form of a matrix product:

$$\begin{bmatrix} \partial x & -\partial y \\ \partial y & \partial x \end{bmatrix} \begin{bmatrix} f(x,y) & g(x,y) \\ -g(x.y) & f(x,y) \end{bmatrix} \tag{4.28}$$

More upon this shortly.

Setting $g(x,y)=0$ gives the gradient:

$$\frac{\partial \begin{bmatrix} f(x,y) & 0 \\ 0 & f(x,y) \end{bmatrix}}{\partial \begin{bmatrix} x & y \\ -y & x \end{bmatrix}} = \begin{bmatrix} \dfrac{\partial f}{\partial x} & -\dfrac{\partial f}{\partial y} \\ \dfrac{\partial f}{\partial y} & \dfrac{\partial f}{\partial x} \end{bmatrix} = grad(f) \tag{4.29}$$

We see that the gradient is a 1-form (conjugate).

Of course, the above assumes that all the derivatives exist at the points of evaluation.

Just for interest, let us differentiate (4.14). We prefer to always specify the real part and the imaginary part, as matrix notation compels us to do, because this avoids any ambiguity regarding the powers of z - is $z^2 = (a+ib)^2$ or is $z^2 = (a+ib)(a-ib)$. We have:

$$\frac{\partial f}{\partial z} = \begin{bmatrix} \partial x & -\partial y \\ \partial y & \partial x \end{bmatrix} \begin{bmatrix} x^3 - 3xy^2 & 3x^2y - y^3 \\ -(3x^2y - y^3) & x^3 - 3xy^2 \end{bmatrix}$$

$$= \begin{bmatrix} (3x^2 - 3y^2) + (3x^2 - 3y^2) & 6xy + 6xy \\ -(6xy + 6xy) & (3x^2 - 3y^2) + (3x^2 - 3y^2) \end{bmatrix}$$

$$(4.30)$$

In this case, we get the same as the conventional derivative (4.15) except for a factor of two.

Within a factor of two, our derivative concurs with the conventional derivative, but our matrix derivative is more general in that we can differentiate any complex function without having to concern ourselves with the Cauchy-Riemann equations. Consider:

$$f = x^3 - 3xy^2 + i(x^2y - y^3)$$

$$\frac{\partial u}{\partial x} = 3x^2 - 3y^2 \qquad \frac{\partial v}{\partial y} = x^2 - 3y^2 \qquad (4.31)$$

$$\frac{\partial u}{\partial x} \neq \frac{\partial v}{\partial y}$$

The Cauchy-Riemann equations of this function, (4.31), are not satisfied and so this function is not differentiable as a conventional Cauchy-Riemann derivative. (Nor is it integrable because it is not differentiable.)

None-the-less, with the more general matrix derivative, we have:

$$f = x^3 - 3xy^2 + i\left(x^2 y - y^3\right)$$

$$\frac{\partial f}{\partial z} = \begin{bmatrix} \partial x & -\partial y \\ \partial y & \partial x \end{bmatrix} \begin{bmatrix} x^3 - 3xy^2 & x^2 y - y^3 \\ -\left(x^2 y - y^3\right) & x^3 - 3xy^2 \end{bmatrix}$$

$$= \begin{bmatrix} \left(3x^2 - 3y^2\right) + \left(x^2 - 3y^2\right) & 2xy + 6xy \\ -\left(2xy + 6xy\right) & \left(3x^2 - 3y^2\right) + \left(x^2 - 3y^2\right) \end{bmatrix}$$

$$(4.32)$$

This is clearly a complex number, and so the function is differentiable using the more general matrix derivative. Put that in your pipe and smoke it!

Reflections upon the Cauchy-Riemann equations:
With thought, we see that the more general matrix differentiation contains within it the conventional differentiation except for a factor of two. Thus all the mathematical structure built upon the conventional differentiation is 'contained' within the general matrix differentiation. So why do we have the Cauchy-Riemann equations?

Looking at (4.18), we see that the Cauchy-Riemann equations are a way of ensuring that the differential is a complex number. Multiplicative closure of the complex numbers requires that the product which is the differential must be an element of the algebra – a complex number.

We have four differentials like (4.16):

$$\frac{\partial u}{\partial x}, \quad \frac{\partial v}{\partial x}, \quad \frac{\partial u}{\partial y}, \quad \frac{\partial v}{\partial y} \qquad (4.33)$$

We need only two differentials to be the real and the imaginary parts of the complex number which is the complex differential. How do we get from four differentials to two differentials? One way is to insist that the four differentials, (4.33), satisfy the Cauchy Riemann equations. Another way is to combine the differentials as:

$$\frac{\partial u}{\partial x}+\frac{\partial v}{\partial y}, \qquad \frac{\partial v}{\partial x}-\frac{\partial u}{\partial y} \qquad\qquad (4.34)$$

This, (4.34), is what the general matrix differentiation does. When the four differentials, (4.33), satisfy the Cauchy-Riemann equations, we get a factor of two emerging. We could have found other such combinations, but that is not what is indicated by (4.21) & (4.27).

The essential difference between the Cauchy-Riemann differentiation and the more general matrix differentiation is the way in which we reduce four differentials to the two parts of a complex number.

Sorry, the conventional Cauchy-Riemann differentiation is perhaps in error; it is certainly of limited application. Perhaps this has implications for conventional integration within the complex numbers.

Differential operators:
It is fashionable within quantum physics to associate differentiation with an operator. Differentiation is differentiation, and we have no need of the concept of operators. Differentiation within the complex numbers ought to be always done as we have done it above, (4.26), but it is cumbersome. There is a shortcut. Because differentiation is a linear operation and matrix multiplication is linear multiplication, we can 'invent' a calculative shortcut:

$$\frac{\partial\begin{bmatrix} f(x,y) & g(x,y) \\ -g(x.y) & f(x,y) \end{bmatrix}}{\partial\begin{bmatrix} x & y \\ -y & x \end{bmatrix}} \equiv \begin{bmatrix} \partial x & -\partial y \\ \partial y & \partial x \end{bmatrix}\begin{bmatrix} f & g \\ -g & f \end{bmatrix} \qquad (4.35)$$

$$= \begin{bmatrix} \dfrac{\partial f}{\partial x}+\dfrac{\partial g}{\partial y} & \dfrac{\partial g}{\partial x}-\dfrac{\partial f}{\partial y} \\ -\left(\dfrac{\partial g}{\partial x}-\dfrac{\partial f}{\partial y}\right) & \dfrac{\partial f}{\partial x}+\dfrac{\partial g}{\partial y} \end{bmatrix} \qquad (4.36)$$

We call the matrix:

$$d = \begin{bmatrix} \partial x & -\partial y \\ \partial y & \partial x \end{bmatrix} \tag{4.37}$$

the differentiation operator. Notice the position of the minus sign in the differentiation operator (4.37). The differential operator is the inverse of the matrix form of an algebra with differentials in place of each variable:

$$\mathbb{C} = \begin{bmatrix} a & b \\ -b & a \end{bmatrix} \qquad \frac{1}{\begin{bmatrix} 0 & 1 \\ -1 & 0 \end{bmatrix}} = \begin{bmatrix} 0 & -1 \\ 1 & 0 \end{bmatrix} \tag{4.38}$$

This definition of the differential operator as the inverse of the usual algebraic form is general for all division algebras. This is because the variable with respect to which we differentiate is a denominator in the differential.

The differentiation operator acts as if by matrix multiplication. Because the complex numbers are commutative, this operator can act from either the right or from the left with the same results:

$$\begin{bmatrix} f & g \\ -g & f \end{bmatrix} \begin{bmatrix} \partial x & -\partial y \\ \partial y & \partial x \end{bmatrix} = \begin{bmatrix} \dfrac{\partial f}{\partial x} + \dfrac{\partial g}{\partial y} & \dfrac{\partial g}{\partial x} - \dfrac{\partial f}{\partial y} \\ -\left(\dfrac{\partial g}{\partial x} - \dfrac{\partial f}{\partial y} \right) & \dfrac{\partial f}{\partial x} + \dfrac{\partial g}{\partial y} \end{bmatrix} \tag{4.39}$$

We can separate this differentiation operator into its real part and its imaginary part:

$$\begin{bmatrix} \partial x & 0 \\ 0 & \partial x \end{bmatrix} \equiv \frac{\partial}{\partial x}, \qquad \begin{bmatrix} 0 & -\partial y \\ \partial y & 0 \end{bmatrix} \equiv -i \frac{\partial}{\partial y} = p_x \tag{4.40}$$

We see the imaginary part is what is called the momentum operator in quantum mechanics. We see that the momentum operator is no more than 'differentiate with respect to minus the imaginary variable'. The energy operator is then differentiate with respect to plus the imaginary variable.

Differentiating a real function with respect to a complex differential using the differential operator gives the gradient:

$$\begin{bmatrix} \partial x & -\partial y \\ \partial y & \partial x \end{bmatrix} \begin{bmatrix} f & 0 \\ 0 & f \end{bmatrix} = \begin{bmatrix} \dfrac{\partial f}{\partial x} & -\dfrac{\partial f}{\partial y} \\ \dfrac{\partial f}{\partial y} & \dfrac{\partial f}{\partial x} \end{bmatrix} = grad(f) \qquad (4.41)$$

Differentiation in 2-dimensional space-time:

There are two 2-dimensional division algebras. There are the complex numbers, $\mathbb{C}$, which are often called the Euclidean complex numbers, and there are the hyperbolic complex numbers, $\mathbb{S}$, which are 2-dimensional space-time. Note that 2-dimensional space-time is a division algebra but that our 4-dimensional space-time is not a division algebra:

$$\mathbb{S} = \exp\left(\begin{bmatrix} t & z \\ z & t \end{bmatrix} \right) = \begin{bmatrix} r & 0 \\ 0 & r \end{bmatrix} \begin{bmatrix} \cosh \chi & \sinh \chi \\ \sinh \chi & \cosh \chi \end{bmatrix} \qquad (4.42)$$

Differentiation within the hyperbolic complex numbers is properly done in a way similar to that shown above for the complex numbers, (4.26), but, being lazy, we use the differentiation operator:

$$d = \begin{bmatrix} \partial a & \partial b \\ \partial b & \partial a \end{bmatrix} \qquad (4.43)$$

We have:

$$\begin{bmatrix} \partial a & \partial b \\ \partial b & \partial a \end{bmatrix} \begin{bmatrix} f(a,b) & g(a,b) \\ g(a,b) & f(a,b) \end{bmatrix} = \begin{bmatrix} \dfrac{\partial f}{\partial a} + \dfrac{\partial g}{\partial b} & \dfrac{\partial g}{\partial a} + \dfrac{\partial f}{\partial b} \\ \dfrac{\partial g}{\partial a} + \dfrac{\partial f}{\partial b} & \dfrac{\partial f}{\partial a} + \dfrac{\partial g}{\partial b} \end{bmatrix} \qquad (4.44)$$

These are the 2-dimensional space-time versions of the divergence and the curl.

We have shown the differentiation using the Cartesian form of the hyperbolic complex numbers. We must be wary because the Cartesian form of this algebra is a division algebra only if we impose restrictions of the form $f(a,b) > g(a,b)$. These restrictions are imposed automatically by the trigonometric functions in (4.42). We will differentiate the polar form:

$$\frac{\partial \begin{bmatrix} e^t \cosh z & e^t \sinh z \\ e^t \sinh z & e^t \cosh z \end{bmatrix}}{\partial \begin{bmatrix} t & z \\ z & t \end{bmatrix}} = \frac{\partial \begin{bmatrix} e^t \cosh z & e^t \sinh z \\ e^t \sinh z & e^t \cosh z \end{bmatrix}}{\partial \begin{bmatrix} t & 0 \\ 0 & t \end{bmatrix}}$$

$$+ \frac{\begin{bmatrix} 0 & 1 \\ 1 & 0 \end{bmatrix} \partial \begin{bmatrix} e^t \cosh z & e^t \sinh z \\ e^t \sinh z & e^t \cosh z \end{bmatrix}}{\partial \begin{bmatrix} z & 0 \\ 0 & z \end{bmatrix}}$$

$$= \begin{bmatrix} te^t \cosh z & te^t \sinh z \\ te^t \sinh z & te^t \cosh z \end{bmatrix} + \begin{bmatrix} e^t \cosh z & e^t \sinh z \\ e^t \sinh z & e^t \cosh z \end{bmatrix} \qquad (4.45)$$

$$= \begin{bmatrix} (1+t)e^t & 0 \\ 0 & (1+t)e^t \end{bmatrix} \begin{bmatrix} \cosh z & \sinh z \\ \sinh z & \cosh z \end{bmatrix}$$

We see again that the differential operator produces the same result:

$$\begin{bmatrix} \partial t & \partial z \\ \partial z & \partial t \end{bmatrix} \begin{bmatrix} e^t \cosh z & e^t \sinh z \\ e^t \sinh z & e^t \cosh z \end{bmatrix}$$

$$= \begin{bmatrix} (1+t)e^t & 0 \\ 0 & (1+t)e^t \end{bmatrix} \begin{bmatrix} \cosh z & \sinh z \\ \sinh z & \cosh z \end{bmatrix} \qquad (4.46)$$

Differentiation of a 3-dimensional division algebra:

Consider the commutative 3-dimensional C_3 division algebra[13]:

$$\exp\left(\begin{bmatrix} a & b & c \\ c & a & b \\ b & c & a \end{bmatrix}\right) \tag{4.47}$$

The differentiation operator within this algebra acting upon a potential is (we use the Cartesian form for ease)[14]:

$$\begin{bmatrix} \partial a & \partial b & \partial c \\ \partial c & \partial a & \partial b \\ \partial b & \partial c & \partial a \end{bmatrix}\begin{bmatrix} u(a,b,c) & v(a,b,c) & w(a,b,c) \\ w(a,b,c) & u(a,b,c) & v(a,b,c) \\ v(a,b,c) & w(a,b,c) & u(a,b,c) \end{bmatrix} \tag{4.48}$$

We have:

$$\begin{bmatrix} \partial a & \partial b & \partial c \\ \partial c & \partial a & \partial b \\ \partial b & \partial c & \partial a \end{bmatrix}\begin{bmatrix} u & v & w \\ w & u & v \\ v & w & u \end{bmatrix} \tag{4.49}$$

$$= \begin{bmatrix} \dfrac{\partial u}{\partial a} + \dfrac{\partial w}{\partial b} + \dfrac{\partial v}{\partial c} & \dfrac{\partial v}{\partial a} + \dfrac{\partial u}{\partial b} + \dfrac{\partial w}{\partial c} & \dfrac{\partial w}{\partial a} + \dfrac{\partial v}{\partial b} + \dfrac{\partial u}{\partial c} \\ \dfrac{\partial w}{\partial a} + \dfrac{\partial v}{\partial b} + \dfrac{\partial u}{\partial c} & \dfrac{\partial u}{\partial a} + \dfrac{\partial w}{\partial b} + \dfrac{\partial v}{\partial c} & \dfrac{\partial v}{\partial a} + \dfrac{\partial u}{\partial b} + \dfrac{\partial w}{\partial c} \\ \dfrac{\partial v}{\partial a} + \dfrac{\partial u}{\partial b} + \dfrac{\partial w}{\partial c} & \dfrac{\partial w}{\partial a} + \dfrac{\partial v}{\partial b} + \dfrac{\partial u}{\partial c} & \dfrac{\partial u}{\partial a} + \dfrac{\partial w}{\partial b} + \dfrac{\partial v}{\partial c} \end{bmatrix} \tag{4.50}$$

$$= \begin{bmatrix} Div & Curl_b & Curl_c \\ Curl_c & Div & Curl_b \\ Curl_b & Curl_c & Div \end{bmatrix}$$

---

[13] See : Dennis Morris : Complex Numbers The Higher Dimensional Forms
[14] The Cartesian form is not a division algebra; only the polar form is the division algebra.

Within this particular spinor space, we see the curl defined in a way which is unusual to our view. This space seemingly plays no part of the physics of our universe.

All differentiation:
The point is that all differentiation can be done with a matrix differentiation operator.

Double differentiation:
It might occur to the reader that we can form a double differentiation operator by squaring the differentiation operator matrix. Your author does not recommend this because your author is a simple fellow and likes things to be clear to his weak mind. It does not work. Differentiation, like Klingon revenge, is a dish best served cold.

The non-commutativity of the differentiation operator:
In general, your weak minded author does not like operators. It is always better to do differentiation properly rather than to use the differential operator shortcut. One reason for this is that the differentiation operator, although looking very much like a commutative complex number is actually non-commutative. This is true of differentiation in general; for example, consider the real number differentiation operator:

$$x^2 \left([\partial x] 2x \right) = 2x^2$$
$$2x \left([\partial x] x^2 \right) = 4x^2$$

(4.51)

We use the differential operator because it is a convenient calculation tool. Remember, you are better off just doing proper differentiation and throwing differential operators into the bin.

Summary:
The form of the differential is the form of the product of a variable and the inverse of a variable. As such the differential can be properly

defined in only division algebras (spinor algebras) because these are the only algebraic structures which hold 'proper' multiplication.

The form of the differential varies from one division algebra to another division algebra as the product varies from one division algebra to another division algebra.

Since all division algebra multiplication is based on matrix multiplication, the form of the differential will be based on matrix multiplication.

Since all division algebra multiplication is based on matrix multiplication, all differentiation can be done with a matrix differential operator. The matrix differential operator is non-commutative even in a commutative algebra – be careful.

Another summary:
Look! the above is controversial stuff. Your author is overthrowing very well established mathematics. There will be trouble. Form your own opinion.

Repetition:
The profundity of what we have done is great. It is deserving of even more repetition.

Using Cauchy-Riemann calculus, we can differentiate and integrate only a limited number of functions within the complex numbers, $\mathbb{C}$.

Using the matrix operator calculus, we can differentiate and integrate any continuous function within the complex numbers, $\mathbb{C}$.

The matrix operator calculus is applicable within any division algebra; it is far more general than the Cauchy-Riemann calculus.

Chapter 5

Non-Commutative Differentiation

To differentiate within a non-commutative division algebra such as the quaternions, we need the E/B product. Commutative differentiation can make no sense within a non-commutative division algebra because the differential is a product of which we have two forms. Do we form the differential as $\delta y \dfrac{1}{\delta x}$ or as $\dfrac{1}{\delta x} \delta y$? These are not equal in a non-commutative division algebra.

Non-commutative differentiation:
Well! reality is the ultimate arbiter. Reality seems to accept that the non-commutative differential of a non-commutative potential, Q_{Pot}, such as a quaternion potential is:

$$E = \frac{1}{2}\left(d_L Q_{Pot} + Q_{Pot} d_R\right)$$

$$B = \frac{1}{2}\left(d_L Q_{Pot} - Q_{Pot} d_R\right)$$

(5.1)

Wherein d_L means act on the left with the differential operator and d_R means act on the right with the differential operator; we use the left chiral quaternions[15] as an example:

---

[15] The left-chiral quaternions are the familiar quaternions of Hamilton. The right-chiral quaternions were discovered by your author and originally foolishly named the anti-quaternions by him.

$$d_L Q^*_{Pot} \equiv \begin{bmatrix} \partial t & -\partial x & -\partial y & -\partial z \\ \partial x & \partial t & \partial z & -\partial y \\ \partial y & -\partial z & \partial t & \partial x \\ \partial z & \partial y & -\partial x & \partial t \end{bmatrix} \begin{bmatrix} \phi & -A_x & -A_y & -A_z \\ A_x & \phi & A_z & -A_y \\ A_y & -A_z & \phi & A_x \\ A_z & A_y & -A_x & \phi \end{bmatrix} \quad (5.2)$$

Note the position of the minus signs in the differential operator. We have it that d_R means act on the right with the differential operator:

$$Q^*_{Pot} d_R \equiv \begin{bmatrix} \phi & -A_x & -A_y & -A_z \\ A_x & \phi & A_z & -A_y \\ A_y & -A_z & \phi & A_x \\ A_z & A_y & -A_x & \phi \end{bmatrix} \begin{bmatrix} \partial t & -\partial x & -\partial y & -\partial z \\ \partial x & \partial t & \partial z & -\partial y \\ \partial y & -\partial z & \partial t & \partial x \\ \partial z & \partial y & -\partial x & \partial t \end{bmatrix} \quad (5.3)$$

Of course the proper way to differentiate is to do it cumbersomely a bit at a time[16] as we did it above for the complex numbers, (4.26). The differential operator gives the same results, is much less cumbersome to use, but hides the underlying mathematics. Your author does not enjoy using the differential operator because it hides the mathematics, but the reader does not enjoy looking upon pages of large matrices, and your author respects the gentle reader's humanity.

This differential, (5.1), together with superimposition which we will deal with in a later chapter, simply gives results in accord with reality. The letters E and the B are chosen because, when non-commutatively differentiating a quaternion potential[17], the E-field becomes the electric field (but see below) and the B-field becomes the magnetic field (but see below). When this non-commutative differentiation is applied to the quaternion Dirac equation, we not only find a solution to the neutrino mass problem but we also see the neutrino field emerge as a chiral (left-handed) super-imposition of the B-fields of the two quaternion algebras[18]. Thus we know why the weak force is left-handed.

---

[16] This is done in tedious detail in : Dennis Morris : The Physics of Empty Space.

[17] Technically, the Maxwell equations derive from the six A_3 algebras and not from the quaternion algebras, but this was not realised until after the fields were first named.

[18] See : Dennis Morris : The Quaternion Dirac Equation

It is the success of the non-commutative differential that drives us to consider the E/B product of the earlier chapter. There is no mathematical justification of the form of the non-commutative differential known to your author.

Differentiating a quaternion:
We really ought to give an example of non-commutative differentiation. We will not present the full matrix but, for presentational ease, we will list the top row elements of the matrix. Since the matrix is a quaternion, the top row elements determine the whole matrix. We choose to differentiate a conjugate left-chiral quaternion. From above, (5.2), we have:

$$d_L\Phi \equiv \begin{bmatrix} \partial t & -\partial x & -\partial y & -\partial z \\ \partial x & \partial t & \partial z & -\partial y \\ \partial y & -\partial z & \partial t & \partial x \\ \partial z & \partial y & -\partial x & \partial t \end{bmatrix} \begin{bmatrix} \phi & -A_x & -A_y & -A_z \\ A_x & \phi & A_z & -A_y \\ A_y & -A_z & \phi & A_x \\ A_z & A_y & -A_x & \phi \end{bmatrix} \quad (5.4)$$

$$d_L\Phi_{[1,1]} = \frac{\partial \phi}{\partial t} - \frac{\partial A_x}{\partial x} - \frac{\partial A_y}{\partial y} - \frac{\partial A_z}{\partial z}$$

$$d_L\Phi_{[1,2]} = -\frac{\partial \phi}{\partial x} - \frac{\partial A_x}{\partial t} - \frac{\partial A_y}{\partial z} + \frac{\partial A_z}{\partial y}$$

$$d_L\Phi_{[1,3]} = -\frac{\partial \phi}{\partial y} + \frac{\partial A_x}{\partial z} - \frac{\partial A_y}{\partial t} - \frac{\partial A_z}{\partial x}$$

$$d_L\Phi_{[1,4]} = -\frac{\partial \phi}{\partial z} - \frac{\partial A_x}{\partial y} + \frac{\partial A_y}{\partial x} - \frac{\partial A_z}{\partial t}$$

(5.5)

From above, (5.3), we have:

$$\Phi d_R \equiv \begin{bmatrix} \phi & -A_x & -A_y & -A_z \\ A_x & \phi & A_z & -A_y \\ A_y & -A_z & \phi & A_x \\ A_z & A_y & -A_x & \phi \end{bmatrix} \begin{bmatrix} \partial t & -\partial x & -\partial y & -\partial z \\ \partial x & \partial t & \partial z & -\partial y \\ \partial y & -\partial z & \partial t & \partial x \\ \partial z & \partial y & -\partial x & \partial t \end{bmatrix} \quad (5.6)$$

$$\Phi d_{R[1,1]} = \frac{\partial \phi}{\partial t} - \frac{\partial A_x}{\partial x} - \frac{\partial A_y}{\partial y} - \frac{\partial A_z}{\partial z}$$

$$\Phi d_{R[1,2]} = -\frac{\partial \phi}{\partial x} - \frac{\partial A_x}{\partial t} + \frac{\partial A_y}{\partial z} - \frac{\partial A_z}{\partial y}$$

$$\Phi d_{R[1,3]} = -\frac{\partial \phi}{\partial y} - \frac{\partial A_x}{\partial z} - \frac{\partial A_y}{\partial t} + \frac{\partial A_z}{\partial x}$$ (5.7)

$$\Phi d_{R[1,4]} = -\frac{\partial \phi}{\partial z} + \frac{\partial A_x}{\partial y} - \frac{\partial A_y}{\partial x} - \frac{\partial A_z}{\partial t}$$

We form the E/B differential as (5.1). The E-field is:

$$E_{[1,1]} = \frac{\partial \phi}{\partial t} - \frac{\partial A_x}{\partial x} - \frac{\partial A_y}{\partial y} - \frac{\partial A_z}{\partial z}$$

$$E_{[1,2]} = -\frac{\partial \phi}{\partial x} - \frac{\partial A_x}{\partial t}$$

$$E_{[1,3]} = -\frac{\partial \phi}{\partial y} - \frac{\partial A_y}{\partial t}$$ (5.8)

$$E_{[1,4]} = -\frac{\partial \phi}{\partial z} - \frac{\partial A_z}{\partial t}$$

The B-field is:

$$B_{[1,1]} = 0$$

$$B_{[1,2]} = -\frac{\partial A_y}{\partial z} + \frac{\partial A_z}{\partial y}$$

$$B_{[1,3]} = \frac{\partial A_x}{\partial z} - \frac{\partial A_z}{\partial x}$$ (5.9)

$$B_{[1,4]} = -\frac{\partial A_x}{\partial y} + \frac{\partial A_y}{\partial x}$$

The non-commutative differential is the two fields which are the E-field and the B-field:

$$Non-Commutative\ \ Differential = \begin{Bmatrix} E - Field \\ B - Field \end{Bmatrix} \qquad (5.10)$$

These equations, (5.8) & (5.9), are exactly the equations which define an electric field and a magnetic field except that they are over quaternion space rather than over our 4-dimensional space-time (see below). We chose to differentiate the conjugate quaternion potential to fit with the arbitrary definition of the electric and the magnetic field.

There are eight Maxwell equations of classical electromagnetism. A second differentiation of the above E/B fields of both left-chiral quaternions and right-chiral quaternions together with superimposition of the resultant fields leads directly to the classical Maxwell equations[19]. A superimposition of the above left-chiral E-field and left-chiral B-field and the right-chiral E-field and right-chiral B-field produces the electromagnetic tensor. (Just add all four fields.[20]) Hence the use of the E and the B to denote these fields. However, there is a ninth Maxwell equation which is not usually included in the list of Maxwell equations.

The Maxwell equations describe how electric and magnetic fields are distributed through the 4-dimensional space-time of our universe. The Maxwell equations are meaningless unless we know how distance within our space-time is defined. The ninth Maxwell equation is:

$$d^2 = t^2 - x^2 - y^2 - z^2 \qquad (5.11)$$

We do not have this equation because we are in quaternion space. The distance function in quaternion space is:

$$d^2 = t^2 + x^2 + y^2 + z^2 \qquad (5.12)$$

The 'quaternion' E/B fields above, (5.8) & (5.9), and the associated 'quaternion Maxwell equations', are distributed over quaternion space

---

[19] See : Dennis Morris : The Physics of Empty Space.
    See also : Dennis Morris : Upon General Relativity.
[20] See : Dennis Morris : The Physics of Empty Space.

not over 4-dimensional space-time. We think the E/B fields above, (5.8) & (5.9), are the fields of the weak force.

Summary:

Non-commutative differentiation works. Reality uses non-commutative differentiation. As such, we must accept it. This implies that we must accept the non-commutative E/B product of the earlier chapter.

Non-commutative differentiation has solved the neutrino mass problem and explained why neutrinos are all left-chiral – the neutrino is the B-field, which is chiral.[21]

Non-commutative differentiation is laborious but simple. We have eased the process by using the differential operators instead of differentiating properly, but this will always work. (It will always work if we make sure the minus signs are in the correct places within the operator.)

Differentiation properly happens within only a division algebra. There are only two types of division algebras; these are commutative and non-commutative. We can now differentiate within any division algebra.

Time for the pub.

We need time in the pub to contemplate what to do with our non-commutative differentials.

---

[21] See : Dennis Morris : The Quaternion Dirac Equation.

Chapter 6

Differentiation of a Quaternion Product

A product of two quaternions is a far more complicated object than a product of two real numbers. There are four independent variables in each quaternion, and so there are sixteen possible products.

If we take the product of a quaternion to be simply, Q_1Q_2, then we can differentiate this product to produce one E-field and one B-field. However, to be comprehensive, we really ought to also take the differential of the product in the other order, Q_2Q_1. This will give a second E-field and a second B-field. Thus, the differential of a quaternion product is four fields.

In this chapter, we will use the E-product and the B-product of the left-chiral quaternions rather than the Q_iQ_j type of product, and we will examine the differential of these products.

The differential of the quaternion non-commutative product:
Take two left-chiral quaternion potentials:

$$\Phi1 = \begin{bmatrix} \phi & A_x & A_y & A_z \\ -A_x & \phi & -A_z & A_y \\ -A_y & A_z & \phi & -A_x \\ -A_z & -A_y & A_x & \phi \end{bmatrix} \quad \Phi2 = \begin{bmatrix} \psi & B_x & B_y & B_z \\ -B_x & \psi & -B_z & B_y \\ -B_y & B_z & \psi & -B_x \\ -B_z & -B_y & B_x & \psi \end{bmatrix}$$

$$(6.1)$$

Wherein all elements are functions of $\{t, x, y, z\}$. We will use the left-chiral quaternion differential operator:

49

$$d = \begin{bmatrix} \partial t & -\partial x & -\partial y & -\partial z \\ \partial x & \partial t & \partial z & -\partial y \\ \partial y & -\partial z & \partial t & \partial x \\ \partial z & \partial y & -\partial x & \partial t \end{bmatrix}$$

(6.2)

We form the E-product and B-product of the two potentials as:

$$E_{\text{Prod}} = \frac{1}{2}(\Phi 1 \Phi 2 + \Phi 2 \Phi 1)$$

$$B_{\text{Prod}} = \frac{1}{2}(\Phi 1 \Phi 2 - \Phi 2 \Phi 1)$$

(6.3)

The differential of the quaternion E-product:

We take the non-commutative differential of the E-product giving the E-field of the E-product and the B-field of the E-product. The E-field of the E-product is:

$$E(E_{\text{Prod}})_{[1,1]} = \frac{\partial(\phi\psi)}{\partial t} - \frac{\partial(A_x B_x)}{\partial t} - \frac{\partial(A_y B_y)}{\partial t} - \frac{\partial(A_z B_z)}{\partial t}$$
$$+ \frac{\partial(\psi A_x)}{\partial x} + \frac{\partial(\psi A_y)}{\partial y} + \frac{\partial(\psi A_z)}{\partial z}$$
$$+ \frac{\partial(\phi B_x)}{\partial x} + \frac{\partial(\phi B_y)}{\partial y} + \frac{\partial(\phi B_z)}{\partial z}$$

(6.4)

We also have:

$$E(E_{\text{Prod}})_{[1,2]} = \frac{\partial(\phi B_x)}{\partial t} + \frac{\partial(\psi A_x)}{\partial t}$$
$$- \frac{\partial(\phi\psi)}{\partial x} + \frac{\partial(A_x B_x)}{\partial x} + \frac{\partial(A_y B_y)}{\partial x} + \frac{\partial(A_z B_z)}{\partial x}$$

(6.5)

$$E\left(E_{\text{Prod}}\right)_{[1,3]} = \frac{\partial\left(\phi B_y\right)}{\partial t} + \frac{\partial\left(\psi A_y\right)}{\partial t}$$

$$-\frac{\partial\left(\phi\psi\right)}{\partial y} + \frac{\partial\left(A_x B_x\right)}{\partial y} + \frac{\partial\left(A_y B_y\right)}{\partial y} + \frac{\partial\left(A_z B_z\right)}{\partial y} \qquad (6.6)$$

$$E\left(E_{\text{Prod}}\right)_{[1,4]} = \frac{\partial\left(\phi B_z\right)}{\partial t} + \frac{\partial\left(\psi A_z\right)}{\partial t}$$

$$-\frac{\partial\left(\phi\psi\right)}{\partial z} + \frac{\partial\left(A_x B_x\right)}{\partial z} + \frac{\partial\left(A_y B_y\right)}{\partial z} + \frac{\partial\left(A_z B_z\right)}{\partial z} \qquad (6.7)$$

The E-field of the E-product contains only ten of the possible sixteen products of individual elements of the two potentials.

The B-field of the E-product is:

$$B\left(E_{\text{Prod}}\right)_{[1,1]} = 0$$

$$B\left(E_{\text{Prod}}\right)_{[1,2]} = -\frac{\partial\left(\phi B_z\right)}{\partial y} - \frac{\partial\left(\psi A_z\right)}{\partial y} + \frac{\partial\left(\psi A_y\right)}{\partial z} + \frac{\partial\left(\phi B_y\right)}{\partial z} \qquad (6.8)$$

$$B\left(E_{\text{Prod}}\right)_{[1,3]} = -\frac{\partial\left(\phi B_x\right)}{\partial z} - \frac{\partial\left(\psi A_x\right)}{\partial z} + \frac{\partial\left(\psi A_z\right)}{\partial x} + \frac{\partial\left(\phi B_z\right)}{\partial x} \qquad (6.9)$$

$$B\left(E_{\text{Prod}}\right)_{[1,4]} = -\frac{\partial\left(\phi B_y\right)}{\partial x} - \frac{\partial\left(\psi A_y\right)}{\partial x} + \frac{\partial\left(\psi A_x\right)}{\partial y} + \frac{\partial\left(\phi B_x\right)}{\partial y} \qquad (6.10)$$

The B-field of the E-product contains only six of the ten products of individual elements within the E-field of the E-product.

Taken together, the E-field of the E-product and the B-field of the E-product contain only ten of the possible sixteen products of individual elements of the two potentials.

The differential of the quaternion B-product:
The E-field of the B-product is:

$$E\left(B_{\text{Prod}}\right)_{[1,1]} = \frac{\partial\left(A_y B_z\right)}{\partial x} - \frac{\partial\left(A_z B_y\right)}{\partial x} + \frac{\partial\left(A_z B_x\right)}{\partial y} - \frac{\partial\left(A_x B_z\right)}{\partial y}$$
$$+ \frac{\partial\left(A_x B_y\right)}{\partial z} - \frac{\partial\left(A_y B_x\right)}{\partial z}$$

(6.11)

$$E\left(B_{\text{Prod}}\right)_{[1,2]} = \frac{\partial\left(A_y B_z\right)}{\partial t} - \frac{\partial\left(A_z B_y\right)}{\partial t}$$

(6.12)

$$E\left(B_{\text{Prod}}\right)_{[1,3]} = \frac{\partial\left(A_z B_x\right)}{\partial t} - \frac{\partial\left(A_x B_z\right)}{\partial t}$$

(6.13)

$$E\left(B_{\text{Prod}}\right)_{[1,3]} = \frac{\partial\left(A_x B_y\right)}{\partial t} - \frac{\partial\left(A_y B_x\right)}{\partial t}$$

(6.14)

The B-field of the B-product is:

$$B\left(E_{\text{Prod}}\right)_{[1,1]} = 0$$

$$B\left(E_{\text{Prod}}\right)_{[1,2]} = \frac{\partial\left(A_y B_x\right)}{\partial y} - \frac{\partial\left(A_x B_y\right)}{\partial y} + \frac{\partial\left(A_z B_x\right)}{\partial z} - \frac{\partial\left(A_x B_z\right)}{\partial z}$$

(6.15)

$$B\left(E_{\text{Prod}}\right)_{[1,3]} = \frac{\partial\left(A_x B_y\right)}{\partial x} - \frac{\partial\left(A_y B_x\right)}{\partial x} + \frac{\partial\left(A_z B_y\right)}{\partial z} - \frac{\partial\left(A_y B_z\right)}{\partial z}$$

(6.16)

$$B\left(E_{\text{Prod}}\right)_{[1,4]} = \frac{\partial\left(A_x B_z\right)}{\partial x} - \frac{\partial\left(A_z B_x\right)}{\partial x} + \frac{\partial\left(A_y B_z\right)}{\partial y} - \frac{\partial\left(A_z B_y\right)}{\partial y}$$

(6.17)

Both the E-field of the B-product and the B-field of the B-product contain only six of the possible sixteen products of individual elements of the two potentials. The six products of individual elements of the two potentials contained in the E-field and the B-field of the B-product are the six products which are absent from the E-field and the B-field of the E-product.

Within the four fields above, we have all sixteen individual elements of the two potentials appearing.

Chapter 7

Superimposition of Differentials

In this chapter, we do not discover how to differentiate within our 4-dimensional space-time. We discover how differentials within division algebra spaces are manifest in our 4-dimensional space-time. We discover what we should do with the differentials, both commutative and non-commutative, that we now know how to calculate within the division algebra spaces.

The $C_2 \times C_2$ algebras:

Remarkably, it seems that the finite group $C_2 \times C_2$ contains all of our known universe except perhaps the strong force. Other than the finite group C_2, no other finite group seems to play any role in our universe.

There are two quaternion algebras which emerge from the finite group $C_2 \times C_2$.[22] These are the left-chiral quaternions and the right-chiral quaternions. Each quaternion has an E-field and a B-field; that is both types of quaternion have a differential.

There are six A_3 algebras which emerge from the finite group $C_2 \times C_2$. Each A_3 algebra has an E-field and a B-field; that is each type of A_3 algebra has a differential.

'Proper' differentiation:

Now, 'proper' differentiation can be done in only a division algebra because only a division algebra has a 'proper' multiplicative product.

---

[22] See : Dennis Morris : The Physics of Empty Space

Great! we understand that, but we do not live in a division algebra space. Our 4-dimensional space-time is not a division algebra space.

Note: If our 4-dimensional space-time was a division algebra space, then we would have the product of norms equal to the norm of the product; this would be:

$$\left(a^2 - b^2 - c^2 - d^2\right)\left(e^2 - f^2 - g^2 - h^2\right) = ?\left(T^2 - X^2 - Y^2 - Z^2\right)$$

$$(7.1)$$

It is a simple algebraic fact that (7.1) is not true. Nor does the distance function of our 4-dimensional space-time emerge as the determinant of any 4-dimensional division algebra.

How do we get from the differentials within division algebra spaces (spinor spaces) to differentials in our universe?

It seems, and we must say "seems" because there is no logical way to prove this, that we simply 'add' the differentials of the different algebraically isomorphic division algebras. Such 'adding' is called superimposition.

The reader might have noticed the inverted commas about the word 'add'. Technically, we can add within only a single division algebra, and so we cannot 'add' the differentials of two different division algebras; even if they are algebraically isomorphic.

Superimposition is adding the matrices of the different algebras.

For example; we can 'add' the left-chiral quaternion matrix to the right-chiral quaternion matrix:

$$\begin{bmatrix} a & b & c & d \\ -b & a & -d & c \\ -c & d & a & -b \\ -d & -c & b & a \end{bmatrix} + \begin{bmatrix} a & b & c & d \\ -b & a & d & -c \\ -c & -d & a & b \\ -d & c & -b & a \end{bmatrix} \qquad (7.2)$$

$$= 2 \begin{bmatrix} a & b & c & d \\ -d & a & 0 & 0 \\ -c & 0 & a & 0 \\ -d & 0 & 0 & a \end{bmatrix} \tag{7.3}$$

This is superimposition; it is not an algebraic operation; in fact, it is mathematical nonsense, but reality seems to accept superimposition.

The superimposed differential:

The superimposed quaternion differential is the matrix 'sum' of the four fields which are the left-chiral quaternion E-field, left-chiral quaternion B-field, right-chiral quaternion E-field, and the right-chiral quaternion B-field.

Using the standard definitions of the electric field and the magnetic field as presented in (5.8) & (5.9):

$$E_t = \frac{\partial \phi}{\partial t} - \frac{\partial A_x}{\partial x} - \frac{\partial A_y}{\partial y} - \frac{\partial A_z}{\partial z} \qquad B_t = 0$$

$$E_x = -\frac{\partial \phi}{\partial x} - \frac{\partial A_x}{\partial t} \qquad B_x = -\frac{\partial A_y}{\partial z} + \frac{\partial A_z}{\partial y}$$

$$E_y = -\frac{\partial \phi}{\partial y} - \frac{\partial A_y}{\partial t} \qquad B_y = \frac{\partial A_x}{\partial z} - \frac{\partial A_z}{\partial x} \tag{7.4}$$

$$E_z = -\frac{\partial \phi}{\partial z} - \frac{\partial A_z}{\partial t} \qquad B_z = -\frac{\partial A_x}{\partial y} + \frac{\partial A_y}{\partial x}$$

We have the superimposed quaternion differential is[23]:

---

[23] This is done in much greater detail in Dennis Morris : The Physics of Empty Space – chapter 15.

$$E_{\mathbb{H}_{Left-Chiral}} + E_{\mathbb{H}_{Right-Chiral}} + B_{\mathbb{H}_{Left-Chiral}} + B_{\mathbb{H}_{Right-Chiral}}$$

$$= \begin{bmatrix} 0 & E_x & E_y & E_z \\ -E_x & 0 & -B_z & B_y \\ -E_y & B_z & 0 & -B_x \\ -E_z & -B_y & B_x & 0 \end{bmatrix} \qquad (7.5)$$

Wherein we have taken the divergence to be zero. This is identical to the standard electromagnetic tensor.

The reader might note that, looking at the distribution of minus signs in the electromagnetic tensor, (7.5), this tensor is 'trying' to be a left-chiral quaternion rather than a right-chiral quaternion. This is intimately connected to the neutrino field being left-chiral.

It seems that the differential of the two quaternion potentials (the left-chiral and the right-chiral) manifest in our 4-dimensional space-time is the superimposition of those two quaternion differentials.

Actually, for simplicity of presentation, we have misled the reader a little. The electromagnetic tensor does not emerge from the superimposition of the two quaternion differentials; we think that (7.5) is the weak force tensor. The electromagnetic tensor emerges as the anti-symmetric part of the superimposition of the differentials of the six A_3 algebras (the sum of six E-fields and six B-fields). The symmetric part of the superimposition of the differentials of the six A_3 algebras is seemingly the energy momentum tensor of general relativity[24].

Second differentials:
If we take the second differentials of the two quaternion potentials and we superimpose those second differentials, we are led to a set of equations which are identical to the Maxwell equations of electromagnetism, both homogeneous and inhomogeneous equations.

---

[24] See : Dennis Morris : Upon General Relativity

The actual Maxwell equations emerge as the anti-symmetric part of the superimposition of the second differentials of the six A_3 algebras.

Electrons and neutrinos:

We really ought to mention that, seemingly, the electron field and the neutrino field emerge from the superimposition of left-chiral quaternions and right-chiral quaternions. The neutrino field is massless, but the neutrino field squared is massive; this is why neutrinos can both travel at the speed of light and oscillate between generations[25].

You see the large amount of physical evidence that superimposition is accepted by the real universe in spite of it being mathematical nonsense.

Summary:

In this chapter, we have not discovered how to differentiate within our 4-dimensional space-time. We have, seemingly, discovered how differentials within division algebra spaces are manifest in our 4-dimensional space-time.

Superimposition seems to be the mechanism by which the differentials within division algebras manifest themselves in our 4-dimensional space-time.

There is a large amount of evidence that the universe accepts superimposition in spite of it being mathematical nonsense.

---

[25] See : Dennis Morris : The Quaternion Dirac Equation
See also: Dennis Morris : The Electron

Chapter 8

Gauge Covariant Differentiation

The division algebra spaces of the complex numbers, $\mathbb{C}$, and the quaternions (both types) are both called gauge spaces and referred to respectively as $U(1)$ and $SU(2)$. There is a third gauge space called $SU(3)$ which is not connected to division algebras in any way.

We know how to differentiate within each of the division algebra gauge spaces, and we know how to form the superimposition of those differentials. In this chapter, we move from the superimposition differential into our 4-dimensional space-time.

Our 4-dimensional space-time:
Our 4-dimensional space-time is a fabrication of six 2-dimensional division algebra spaces with four copies of the 1-dimensional division algebra that is the real numbers. Three of the 2-dimensional spaces are the 2-dimensional Euclidean spaces of the complex numbers (the complex plane); the other three of the 2-dimensional spaces are the 2-dimensional hyperbolic complex numbers which are 2-dimensional space-time.

To put it simply, our 4-dimensional space-time is a fabrication of six 2-dimensional planes together with the four copies of the real numbers. We know this because we observe 2-dimensional angles within our 4-dimensional space-time and only 2-dimensional division algebra spaces have 2-dimensional angles.

Our 4-dimensional space-time is an emergent expectation space which emerges from the superimposition of the six A_3 algebras[26]; it is not a division algebra space.

---

[26] See : Dennis Morris : Upon General Relativity

The differential is a product. The 'proper[27]' multiplication operation exists in only division algebras. So how do you differentiate in a space that is not a division algebra space? We will address that question in a later chapter; in this chapter, we concentrate on something different.

Gauge covariant differentiation - explanation:

Our 4-dimensional space-time is not the only emergent expectation space in which we live. It seems that we also inhabit the quaternion emergent expectation space which emerges from the superimposition of the two quaternion algebras[28]. This space manifests itself as a gauge space forming part of a fibre bundle (more explanation later) with 4-dimensional space-time as the basis space to which the quaternion space is attached from point to point. The quaternion spaces seem to be associated with the electroweak force.

Aside: It seems that there is a third emergent expectation space associated with the strong nuclear force. We speculate that this is a folded 8-dimensional emergent space, but we will not consider this space because we understand very little about it. In the standard model, the gauge space $SU(3)$ plays the role of this third emergent expectation space.

Fibre bundles:

There is a fundamental difference between the way Isaac Newton viewed space and time and the way that we view space and time today. Newton thought of time as a 1-dimensional space completely separate from the 3-dimensional space we see around us. He took the view that there was a copy of our 3-dimensional space fixed to every point of the 1-dimensional space that is time. Newton viewed space and time as what mathematicians call a fibre bundle.

---

[27] 'Proper' means satisfies the axioms of (has the properties of) a division algebra like multiplicative closure, multiplicative inverses, absence of zero divisors etc..
[28] See : Dennis Morris : The Electron

We illustrate Newton's view with a 2-dimensional plane instead of the 3-dimensional space:

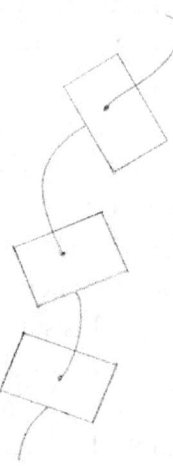

A fibre bundle is a few (could be one or many) separate types of space fixed on to an underlying space at every point of that underlying space. There is nothing that says the two or more spaces that are tied together in a fibre bundle have to be of the same dimension or of any particular dimension.

Our modern view is that 3-dimensional space and 1-dimensional time are not two separate spaces but together form a single 4-dimensional space called 4-dimensional space-time.

This means that, since the four dimensions are in the same space, we can rotate in a space-time plane as well as rotating in a purely spatial plane. This ability to rotate in any plane formed of any two axes, including the time axis, is exactly what we mean by saying our 4-dimensional space-time is a single space rather than two separate spaces tied together as a fibre bundle. We cannot form a 2-dimensional plane from two axes that are in separate spaces.

So, to restate the question, what is differentiation in a fibre bundle of emergent expectation spaces? Remarkably, it seems that we know the answer.

That answer is that, basically,

a) We differentiate a potential within each copy of the division algebra space which is the gauge space.
b) We take the superimposition of the differentials of the isomorphic gauge spaces (algebraically isomorphic division algebras) by simply 'adding' the differentials.
c) We take the superimposition of the potentials within the isomorphic gauge spaces by simply 'adding' the potentials.
d) We add or subtract a bit of the superimposed potential to adjust the superimposed differential so that it fits into our 4-dimensional space-time properly.

I will re-emphasize the important bit. We add or subtract a bit of the superimposed gauge potential to our superimposed gauge differential to make it fit into our 4-dimensional space-time properly.

Firstly:

We are going to differentiate within the complex numbers. Because there is only one copy of the complex numbers which emerges from the finite group C_2, we do not need to differentiate in more than one copy of the complex numbers and superimpose the results to get our gauge space differential.

If we were differentiating within the quaternions, because there are two copies of the quaternion algebras which emerge from the finite group $C_2 \times C_2$, the left-chiral quaternions and the right-chiral quaternions[29], we would get our gauge space differential by differentiating within both algebras and superimposing (adding) the results.

Having obtained the superimposed gauge space differential, we need to find how the superimposed differential of that gauge space is manifest in our 4-dimensional space-time. The resulting differential in our 4-dimensional space-time, is called the gauge covariant differential. It should really be called the superimposed gauge covariant differential.

---

[29] See : Dennis Morris : Quaternions

Finding the (superimposed) gauge covariant differential is what will occupy us for this chapter.

For now, since we are working with the complex numbers, we do not worry about superimposition of the differentials; we just concentrate upon the covariant part of the differentiation.

QED:

Within quantum electro dynamics, we have the concept of the Lie group $U(1)$. Although called a Lie group, $U(1)$, is really the unit circle in the complex plane. Within QED, we speak of locally varying $U(1)$ phase over 4-dimensional space-time.

Aside: We speak also of locally varying $SU(2)$ phase over 4-dimensional space-time, and similarly of locally varying $SU(3)$ phase, but we will ignore these more complicated Lie groups for now. $SU(2)$ is connected to the quaternions.

$U(1)$ phase is just another phrase for angle subtended at the origin between the real axis and a point in the complex plane.

The central idea of QED is that we have a base space which is our 4-dimensional space-time and that we have also three gauge spaces, $\{U(1), SU(2), SU(3)\}$, which are attached to our 4-dimensional space-time at each point of that 4-dimensional space-time.[30] For the present, we will consider only the $U(1)$ gauge space (the complex numbers, $\mathbb{C}$,).

Thus, the QED view is that, at each point in our 4-dimensional space-time, we have an attached copy of the complex plane. Imagine pinning a complex plane through the origin to every point of our space-time. Actually, it is sufficient to imagine pinning through the origin copies of the complex plane to only two separate points of our 4-dimensional space-time. Now imagine spinning the two copies of the complex plane

---

[30] For a discussion about how these gauge spaces are attached to our 4-dimensional space-time, see : Dennis Morris : The Electron.

like a roulette wheel and waiting until they stop[31]. It is unlikely that the two copies of the complex plane will come to rest with their real axis pointing in the same direction.

Parallel transport:

Within our 4-dimensional space-time, we can move a vector (a little arrow) from point to point in such a way that the vector points in the same direction at the different points. We say that we have parallel transported the vector from point A to point B in our space-time if the direction pointed by the vector is unchanged by the movement. Thus we know what parallel means in our 4-dimensional space-time. It is against this that we measure the direction of the real axis of the complex planes.

However, this vector exists not in only our space-time but also in the gauge space which is the complex plane. Perhaps the orientation of the real axis of the complex plane is not parallel at the different points A and B of our space-time. We have a conflict over the meaning of parallel between the base space, which is our 4-dimensional space-time, and the gauge space, which is the complex plane, because the axes of the two separate copies of the complex plane are not aligned. The vector cannot be parallel in both the gauge space and our 4-dimensional space-time at once (unless, by some coincidence, the real axes of the two copies of the complex plane are aligned as measured against parallel transport in our space-time).

A picture is worth a thousand words. We have one worth a few dozen words.

---

[31] Okay, spinning space in space has no friction, and so it will not stop, but you get the idea.

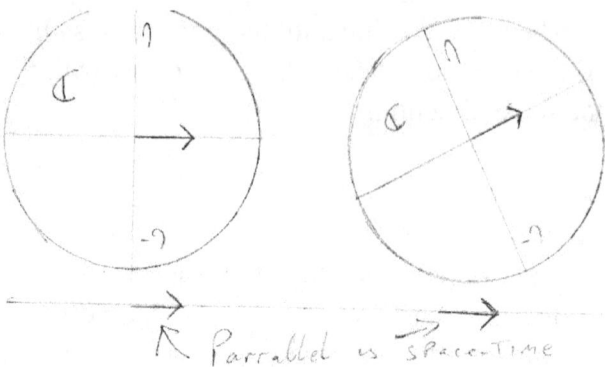

We see that the same vector is parallel with itself at different points in our 4-dimensional space-time; it is also parallel with itself at different points in the gauge space as measured by the axes of that gauge space. However, because the axes of the complex plane are not tied together, the vector in the gauge space does not appear to be parallel to itself when viewed from 4-dimensional space-time. We have a conflict of affine connections.

Note that we could take the view that the arbitrarily defined axes of the complex planes are orientated in the same way at both points in our 4-dimensional space-time but that the point in the complex plane which is the tip of the vector has moved within the complex plane. Such a change would be a complex function of the two space-time variables. Remember that a complex function is a potential – see (4.27).

Well, above we have outlined the central idea of QED. How does the maths work?

Gauge covariant differentiation - the maths:
We begin with a gauge potential, Φ, and we wish to differentiate that gauge potential. This differentiation can be done only if the gauge potential is a division algebra such as the quaternions (left-chiral or right–chiral) or the complex numbers or some other division algebra. We are going to choose a complex number gauge potential, $\Phi(a,b)$:

$$\Phi(a,b) = \begin{bmatrix} \phi(a,b) & A_x(a,b) \\ -A_x(a,b) & \phi(a,b) \end{bmatrix} \tag{8.1}$$

This is constructing what physicists call a $U(1)$ gauge theory. The variables, (a,b), are two of the four variables of our 4-dimensional space-time; thus the complex number potential is a complex number which varies from point to point in a 2-dimensional plane within our 4-dimensional space-time. Of course, a complex number is a pair of real numbers and we can think of it as a 2-dimensional vector.

Great! We know how to differentiate within the complex numbers, we just apply the differentiation operator:

$$\begin{bmatrix} \partial a & -\partial b \\ \partial b & \partial a \end{bmatrix} \begin{bmatrix} \phi(a,b) & A_x(a,b) \\ -A_x(a,b) & \phi(a,b) \end{bmatrix} = \begin{bmatrix} \dfrac{\partial \phi}{\partial a} + \dfrac{\partial A_x}{\partial b} & \dfrac{\partial A_x}{\partial a} - \dfrac{\partial \phi}{\partial b} \\ -\left(\dfrac{\partial A_x}{\partial a} - \dfrac{\partial \phi}{\partial b} \right) & \dfrac{\partial \phi}{\partial a} + \dfrac{\partial A_x}{\partial b} \end{bmatrix}$$

$$(8.2)$$

Possibly the most important concept within physics is the idea that things work the same regardless of the direction in space in which the physical apparatus is pointing. Kettles boil at the same temperature when the spout is pointing north as they do when the spout is pointing east or west or south or … This means that we should be able to rotate our potential and differentiate it and get the same result as we do when we differentiate our potential and then rotate it.

How do we rotate a complex number gauge potential? We use a complex number rotation matrix:

$$e^{i\theta}\Phi = \begin{bmatrix} \cos\theta & \sin\theta \\ -\sin\theta & \cos\theta \end{bmatrix} \begin{bmatrix} \phi(a,b) & A_x(a,b) \\ -A_x(a,b) & \phi(a,b) \end{bmatrix}$$

$$= \begin{bmatrix} \phi\cos\theta - A_x\sin\theta & \phi\sin\theta + A_x\cos\theta \\ -(\phi\sin\theta + A_x\cos\theta) & \phi\cos\theta - A_x\sin\theta \end{bmatrix} \tag{8.3}$$

How do we rotate the differential of a potential? We use a complex number rotation matrix:

$$e^{i\theta}\left(\partial_{\mu}\Phi\right)=\begin{bmatrix}\cos\theta & \sin\theta \\ -\sin\theta & \cos\theta\end{bmatrix}\begin{bmatrix}\dfrac{\partial\phi}{\partial a}+\dfrac{\partial A_{x}}{\partial b} & \dfrac{\partial A_{x}}{\partial a}-\dfrac{\partial\phi}{\partial b} \\ -\left(\dfrac{\partial A_{x}}{\partial a}-\dfrac{\partial\phi}{\partial b}\right) & \dfrac{\partial\phi}{\partial a}+\dfrac{\partial A_{x}}{\partial b}\end{bmatrix}$$

$$=\begin{bmatrix}\left(\dfrac{\partial\phi}{\partial a}+\dfrac{\partial A_{x}}{\partial b}\right)\cos\theta-\left(\dfrac{\partial A_{x}}{\partial a}-\dfrac{\partial\phi}{\partial b}\right)\sin\theta & \sim \\ -\left(\dfrac{\partial\phi}{\partial a}+\dfrac{\partial A_{x}}{\partial b}\right)\sin\theta-\left(\dfrac{\partial A_{x}}{\partial a}-\dfrac{\partial\phi}{\partial b}\right)\cos\theta & \sim\end{bmatrix}$$

$$(8.4)$$

Wherein we have squiggled out the duplicate information for presentational ease.

Firstly, we take the rotation to be global. This means that we have rotated the complex gauge potential by the same angle throughout the whole of the 4-dimensional space-time universe. This means that the angle of rotation, θ, does not vary from point; this means that θ is not a function of (a,b). This means that a vector is parallel with itself in both our 4-dimensional space-time and the complex plane which is the gauge space. We apply the differentiation operator to the rotated potential:

$$\partial_{\mu}\left(e^{i\theta}\Phi\right)=\begin{bmatrix}\partial a & -\partial b \\ \partial b & \partial a\end{bmatrix}\begin{bmatrix}\phi\cos\theta-A_{x}\sin\theta & \phi\sin\theta+A_{x}\cos\theta \\ -(\phi\sin\theta+A_{x}\cos\theta) & \phi\cos\theta-A_{x}\sin\theta\end{bmatrix}$$

$$=\begin{bmatrix}\left(\dfrac{\partial\phi}{\partial a}+\dfrac{\partial A_{x}}{\partial b}\right)\cos\theta+\left(\dfrac{\partial\phi}{\partial b}-\dfrac{\partial A_{x}}{\partial a}\right)\sin\theta & \sim \\ -\left(\dfrac{\partial\phi}{\partial a}+\dfrac{\partial A_{x}}{\partial b}\right)\sin\theta-\left(\dfrac{\partial A_{x}}{\partial a}-\dfrac{\partial\phi}{\partial b}\right)\cos\theta & \sim\end{bmatrix}$$

$$(8.5)$$

We see that the result (8.5) is the same as (8.4). This result can be put more simply. The differential operator and the rotation matrix commute for a global rotation.

Now let us consider a rotation which varies locally; this means the rotation varies from point to point in the 2-dimensional plane of our 4-dimensional space-time. This means that a vector which is parallel with itself in our 4-dimensional space-time is not parallel with itself in the complex plane. In this case, the angle of rotation of the axes of the complex plane, $\theta(a,b)$, is a function of the space-time co-ordinates.

Rotating the differential gives the same result for this local rotation as for the global rotation, (8.4).

Differentiating the rotated potential gives:

$$\partial_\mu \left(e^{i\theta}\Phi \right) = \begin{bmatrix} \partial a & -\partial b \\ \partial b & \partial a \end{bmatrix} \begin{bmatrix} \phi\cos\theta - A_x\sin\theta & \phi\sin\theta + A_x\cos\theta \\ -\phi\sin\theta - A_x\cos\theta & \phi\cos\theta - A_x\sin\theta \end{bmatrix}$$

$$= \begin{bmatrix} \left(\begin{bmatrix} \left(\dfrac{\partial\phi}{\partial a} + \dfrac{\partial A_x}{\partial b}\right)\cos\theta + \left(\dfrac{\partial\phi}{\partial b} - \dfrac{\partial A_x}{\partial a}\right)\sin\theta \\ + \left[\phi\left(\dfrac{\partial\cos\theta}{\partial a} + \dfrac{\partial\sin\theta}{\partial b}\right) + A_x\left(\dfrac{\partial\cos\theta}{\partial b} - \dfrac{\partial\sin\theta}{\partial a}\right)\right] \end{bmatrix}\right) & \sim \\ \left(\begin{bmatrix} -\left(\dfrac{\partial\phi}{\partial a} + \dfrac{\partial A_x}{\partial b}\right)\sin\theta - \left(\dfrac{\partial A_x}{\partial a} - \dfrac{\partial\phi}{\partial b}\right)\cos\theta \\ + \left[\phi\left(\dfrac{\partial\cos\theta}{\partial b} - \dfrac{\partial\sin\theta}{\partial a}\right) - A_x\left(\dfrac{\partial\cos\theta}{\partial a} + \dfrac{\partial\sin\theta}{\partial b}\right)\right] \end{bmatrix}\right) & \sim \end{bmatrix}$$

$$(8.6)$$

We see that, because of the product rule of differentiation, we have extra terms in this expression, (8.6), compared with (8.4). For a local rotation, the differential operator and the rotation matrix do not commute. The extra terms are the complex number:

$$\begin{bmatrix} \phi\left(\dfrac{\partial\cos\theta}{\partial a} + \dfrac{\partial\sin\theta}{\partial b}\right) + A_x\left(\dfrac{\partial\cos\theta}{\partial b} - \dfrac{\partial\sin\theta}{\partial a}\right) & \sim \\ \phi\left(\dfrac{\partial\cos\theta}{\partial b} - \dfrac{\partial\sin\theta}{\partial a}\right) - A_x\left(\dfrac{\partial\cos\theta}{\partial a} + \dfrac{\partial\sin\theta}{\partial b}\right) & \sim \end{bmatrix} \quad (8.7)$$

$$= \begin{bmatrix} \phi & A_x \\ -A_x & \phi \end{bmatrix} \cdot \left(\begin{bmatrix} \partial a & -\partial b \\ \partial b & \partial a \end{bmatrix}\begin{bmatrix} \cos\theta & \sin\theta \\ -\sin\theta & \cos\theta \end{bmatrix}\right) \quad (8.8)$$

Note that we have to put the differentiation operator in a bracket with the matrix upon which it is operating for notational clarity.

Looking at (8.8), we see the extra term complex number is a multiple of the potential. The extra bit of potential is of the form:

$$\begin{bmatrix} x & y \\ -y & x \end{bmatrix}\begin{bmatrix} \phi & A_x \\ -A_x & \phi \end{bmatrix} = \mathbb{C}.\Phi \tag{8.9}$$

The photon field:

In QED, the negative of the extra bit of potential, (8.8), which has arisen as a consequence of local rotation of the attached complex plane – that's locally varying $U(1)$ phase – is the electromagnetic photon field. Photons exist because the orientations of the axes of the copies of the complex plane that are pinned through the origin to separate points in our space-time vary from point to point over our space-time.

We'll say that again; photons exist because the copies of the complex plane have axes which are orientated differently at each point in our space-time.

If you prefer, photons exist because there is a complex function (a potential which varies from point to point in our 4-dimensional space-time – see (4.27).

Interim summary:

So far this chapter is quite a shock to the reader who has not met QED and gauge covariant differentiation before. We are asserting that we live in several types of space rather than only 4-dimensional space-time. We are asserting that our 4-dimensional space-time is a base space (in a fibre bundle) to which is attached (pinned through the origin if you like) a copy of each of the other (gauge) spaces at every point in our 4-dimensional space-time. These other pinned spaces are called gauge spaces. Above we have considered only one of these gauge spaces, the 2-dimensional complex plane, and we have found that an 'extra bit' of potential arises as a consequence of the orientation of the axes in the separate copies of the complex plane being out of alignment as measured against a parallel transported vector in 4-dimensional space-time.

Covariant differentiation in gauge theory:
So what is gauge covariant differentiation? Gauge covariant differentiation is just differentiation with an extra term. The extra term just subtracts the extra potential. We signify the gauge covariant differential by D or D_μ rather than by ∂ or ∂_μ. We have the $U(1)$ gauge covariant differential as:

$$D_\mu^{U(1)} \begin{bmatrix} \phi & A_x \\ -A_x & \phi \end{bmatrix} = \begin{bmatrix} \partial a & -\partial b \\ \partial b & \partial a \end{bmatrix} \begin{bmatrix} \phi & A_x \\ -A_x & \phi \end{bmatrix}$$
$$-\begin{bmatrix} \phi & A_x \\ -A_x & \phi \end{bmatrix} \left(\begin{bmatrix} \partial a & -\partial b \\ \partial b & \partial a \end{bmatrix} \begin{bmatrix} \cos\theta & \sin\theta \\ -\sin\theta & \cos\theta \end{bmatrix} \right)$$

$$(8.10)$$

Using the covariant differential, the differential of a rotated potential is equal to a rotated differentiated potential:

$$e^{i\theta} \left(D_\mu \Phi \right) = D_\mu e^{i\theta} \Phi \qquad (8.11)$$

We have resolved the conflict between parallel and not-parallel by simply adjusting the differential by subtracting the 'unwanted bit'. Gauge covariant differentiation – big name – simple subtraction.

The reader is unlikely to see the gauge covariant differential presented as matrices elsewhere in the literature. Most often, the reader will see the gauge covariant derivative presented as:

$$D_\mu \varphi = \partial_\mu \varphi - iB_\mu \varphi \qquad (8.12)$$

Unfortunately, we have a notational lack of clarity here, (8.12), in that the potential is denoted in (8.12) by φ and the complex number multiplying that potential is denoted by iB_μ.

What about the other 2-dimensional planes?:

There are three copies of the complex plane in our 4-dimensional space-time, but every combination of 2-dimensional Euclidean rotations in our 4-dimensional space-time is a single 2-dimensional Euclidean rotation. We therefore need to take no account of more than one $U(1)$ gauge space. The same is true of the hyperbolic complex numbers which are 2-dimensional space-time.

Other gauge covariant differentials:

The standard model of particle physics has three gauge spaces. These are the complex plane, $U(1)$, the quaternion space, $SU(2)$[32], and the $SU(3)$ Lie group which is not associated with a division algebra space[33]. Each gauge space has its own gauge covariant derivative based upon the normal derivative within the division algebra which is the gauge space.

The quaternion covariant derivative:

Clearly, for the quaternions, the gauge covariant derivative will be non-commutative and we will have E-fields and B-fields. More importantly, there are two quaternion algebras, the left-chiral quaternions and the right-chiral quaternions. We form the gauge derivative by taking the differential of the quaternion potential in each of these two algebras and then superimposing ('adding') the two differentials ('adding' the matrices of the E-fields and the B-fields).

Such superimposition smashes the algebraic structure of the quaternions. It seems to smash the algebraic structure into three copies of the 2-dimensional complex numbers, $\mathbb{C}$, but we are unsure of this.

---

[32] There are two quaternion division algebras, the left-chiral and the right-chiral, but the standard model uses only one and does not specify which it uses.

[33] Your author does not like $SU(3)$.

We then add/subtract an extra term (extra bit of the potential) to this sum of differentials to form the quaternion, $SU(2)$, gauge covariant derivative.

The superimposition of the two quaternion differentials is not a quaternion, and so the extra term (extra bit of the potential) in the covariant derivative will not be a quaternion.

Because quaternion space is 4-dimensional rather than 2-dimensional, there are actually three extra terms (extra bits of the potential corresponding to the three quaternion imaginary variables) added/subtracted from the gauge derivative to fit with the three Euclidean rotation planes of our 4-dimensional space-time.

As the reader might imagine, the resulting gauge covariant derivative is a complicated object, and we do not wish to present it to the reader in this small book.

The hyperbolic complex numbers covariant derivative:
There are two types of 2-dimensional division algebras within our 4-dimensional space-time. We have developed the $U(1)$ complex numbers, $\mathbb{C}$, gauge covariant derivative above, but what about the gauge covariant derivative of the other 2-dimensional division algebra which is the hyperbolic complex numbers?

The orientation of the axes of the hyperbolic complex numbers (the phase) can vary from point to point over our 4-dimensional space-time just as the orientation of the axes of the Euclidean complex numbers can vary from point to point over our 4-dimensional space-time. What is variation of orientation of axes of the hyperbolic complex numbers, variation of hyperbolic phase if you prefer? It is variation of velocity. A rotation in the hyperbolic complex numbers is just a change of velocity. Local variation of axes orientation is just local variation of velocity – acceleration.

Acceleration is associated with a force and with an amount of mass. This is very similar to local variation of $U(1)$ phase (the Euclidean complex numbers) being associated with an electromagnetic force and an amount of electric charge.

Now consider two copies of the hyperbolic complex numbers pinned through the origin to two different points, A & B, in our space-time. Assume the orientation of the axes of the two copies of these hyperbolic complex numbers is not the same as measured against a parallel transported vector in 4-dimensional space-time. Imagine a particle which moves from A to B. As the particle moves, it changes velocity because rotation in the hyperbolic complex numbers is change of velocity.

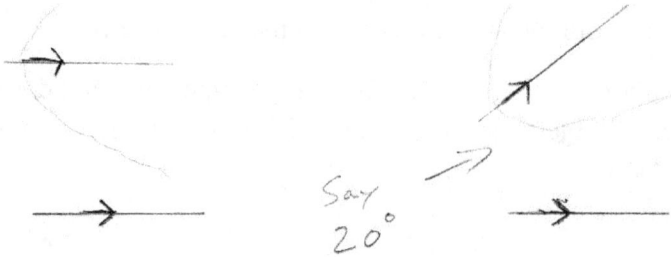

Perhaps the difference in orientation of the axes is hyperbolic 20^0.

Now suppose that the underlying 4-dimensional space-time is suddenly curved by some great being but that the orientations of the axes of the hyperbolic complex number spaces is unchanged.

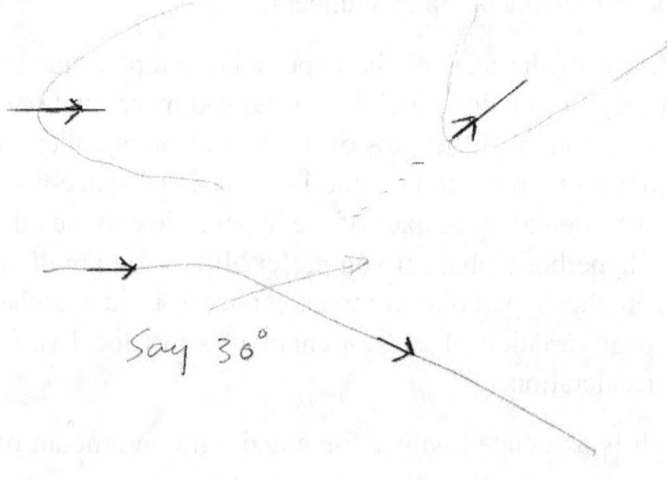

We see that, as a result of the curvature of the underlying 4-dimensional space-time, the difference in orientation of the axes is changed to, say, hyperbolic 30^0. This corresponds to a higher velocity than in the not curved 4-dimensional space-time. The curvature of our 4-dimensional space-time has 'created' an amount of acceleration. This is gravity. The equivalence between inertial mass and gravitational mass is the equivalence of space-time angles measured between two affine connections.

Summary:
Gauge covariant derivatives differ from normal superimposed derivatives by the subtraction of an extra bit of the superimposed potential.

Chapter 9

Covariant Differentiation – General Relativity

There is a second type of covariant derivative.

In the last chapter, we looked at differentiation within a gauge space of which copies are attached, pinned through the origin, to our 4-dimensional space-time at each point of our 4-dimensional space-time.

In this chapter, we look at differentiation within our 4-dimensional space-time. Our 4-dimensional space-time is not a 4-dimensional division algebra space, and so we cannot form a 4-dimensional product. If we cannot form a product, we cannot form a 4-dimensional differential like:

$$\frac{1}{dx}dy \tag{9.1}$$

So, how do we differentiate in our 4-dimensional space-time?

Differentiation in our 4-dimensional space-time:

In short, within our space-time, there are three types of division algebras.

We have both types of 2-dimensional division algebras the complex numbers, $\mathbb{C}$, and the hyperbolic complex numbers, $\mathbb{S}$; these appear as gauge spaces attached to out 4-dimensional space-time as described in the chapter on gauge covariant differentiation. Thus, differentiation within these spaces is gauge covariant differentiation.

We have four copies of the 1-dimensional real numbers. We are going to consider differentiation within the four copies of the real numbers in this chapter. Basically, we differentiate in each independent copy of the real numbers, and we fix the differentials together as a 4-vector.

Thus, the differential in 4-dimensional space-time is four 1-dimensional differentials fitted together as the four components of a 4-vector. This is standard conventional stuff; there is nothing revolutionary here.

More explanation:

Our 4-dimensional space-time is the emergent expectation space of the (six off) A_3 algebras[34].

Our 4-dimensional space-time is a fabrication of six 2-dimensional division algebra spaces. Three of these are copies of the 2-dimensional complex plane, and three of these are copies of the 2-dimensional hyperbolic complex numbers.[35] We have dealt with differentiation within these division algebra spaces above when we did gauge covariant differentiation.

Our 4-dimensional space-time is also a fabrication of four copies of the 1-dimensional real numbers. One of the consequences of the superimposition of the six 4-dimensional A_3 algebras which leads to the emergence of our 4-dimensional space-time is that all algebraic structure is lost in superimposition other than the algebraic structure of the 1-dimensional real numbers. Since we have to differentiate within only division algebra sub-spaces, we differentiate in each of the four 1-dimensional real number spaces and add these differentials as if they were vectors, but there is a complication.

Curvature of 4-dimensional space-time:

If we try to introduce some kind of spatial curvature into the 2-dimensional complex plane, the whole algebraic structure of the complex numbers falls to bits. The same is true of all division algebra spaces. The division algebra spaces (spinor spaces) are all flat. In fact,

---

[34] See : Dennis Morris : Upon General Relativity
[35] For details of how these division algebra spaces become a fabricated space, see : Dennis Morris : Upon General Relativity.

the division algebra spaces are so flat that the concept of zero curvature does not exist within them.

Because our 4-dimensional space-time is not a division algebra space, it is not held rigidly flat. Within a space which is not a division algebra space, there is no algebraic structure to keep the space flat.

Looking back at the summary of the chapter on gauge covariant differentiation, we see that gauge covariant derivatives differ from normal derivatives by the subtraction of an extra bit of potential. Well, covariant derivatives within our curved 4-dimensional space-time differ from normal derivatives by the subtraction/addition of an extra bit. The extra bit is a measure of the curvature of the space rather than an extra bit of potential.

Not identical twins:
It is confusing that both gauge covariant differentiation and 4-dimensional space-time covariant differentiation are both called covariant differentiation. Since, in both cases, we form the covariant derivative by adding/subtracting a bit of something to the differential, we can see why these very different types of differentiation share the same name.

In spite of the two types of covariant derivative sharing the same name, they are different types of differentiation because the bits we add to the differentials have different natures.

Gauge covariant differentiation is differentiation within the gauge space attached to our 4-dimensional space-time as part of a fibre bundle. 4-dimensional space-time covariant differentiation is differentiation within our 4-dimensional space-time.

The 4-dimensional space-time covariant derivative:
I will draw for the gentle reader a set of straight lines in curved space. So deserving and appreciative of fine art is the wise reader that I will also draw a set of curved lines in flat (not curved) space. The reader

will then be able to understand the difference between these two concepts. We have:

I leave the reader to determine which is which.

The point of the above art is that curved co-ordinates in a flat space are the same as straight co-ordinates in a curved space. The calculus of curved space is simply the calculus of flat space with a curved co-ordinate system. That is worth repeating.

The calculus of curved space is simply the calculus of flat space with a curved co-ordinate system.

Shall we repeat this again? Why not?

The calculus of curved space is simply the calculus of flat space with a curved co-ordinate system. Of course, the curved co-ordinate system need not be a simply or regularly curved co-ordinate system. It is most often a really weirdly curved co-ordinate system, but the truth remains. Once the reader has digested this truth, she will be well on the way to understanding the mathematics of general relativity, and so we will repeat that truth.

The calculus of curved space is simply the calculus of flat space with a curved co-ordinate system.

Differentiating vectors:
Differentiation in our 4-dimensional space-time is differentiation of vectors (sums of 1-dimensional differentiations). For simplicity, we will work in 2-dimensional $\mathbb{R}^2$ space.

Within 2-dimensional flat space with a Cartesian co-ordinate system, a 2-dimensional vector, $\vec{V}$, is of the form:

$$\vec{V} = f(x,y)\vec{e_x} + g(x,y)\vec{e_y} \qquad (9.2)$$

This is $\mathbb{R}^2$ space. Each of the functions, $\{f(x,y), g(x,y)\}$ is a real number; these functions are completely independent, and there is no relationship between these real numbers like $g^2 = -f$. This is not a division algebra space. We are to do 1-dimensional differentiation in each of the separate and independent real number algebras.

The basis vectors, $\vec{e_i}$, in Cartesian co-ordinates are constant; they do not change from point to point in the underlying space. This is a property of the Cartesian co-ordinate system. This contrasts with, say, polar co-ordinates where the radial basis vector and the angular basis vector do vary from point to point in the underlying space.

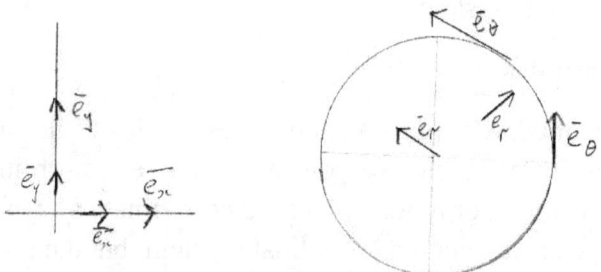

Differentiating a vector in Cartesian co-ordinates (with constant basis vectors) gives (differentiation of a product rule):

$$\frac{\partial}{\partial x}\left(f(x,y)\vec{e_x} + g(x,y)\vec{e_y}\right) = \frac{\partial f}{\partial x}\vec{e_x} + \frac{\partial \vec{e_x}}{\partial x}f + \frac{\partial g}{\partial x}\vec{e_y} + \frac{\partial \vec{e_y}}{\partial x}g$$
$$= \frac{\partial f}{\partial x}\vec{e_x} + 0 + \frac{\partial g}{\partial x}\vec{e_y} + 0 \qquad (9.3)$$

With a similar result for differentiation with respect to y. We see that, because the basis vectors do not vary, their differentials are zero, and so they are traditionally ignored:

$$\frac{\partial \vec{e_x}}{\partial x} = 0, \qquad \frac{\partial \vec{e_y}}{\partial x} = 0 \tag{9.4}$$

Differentiating a vector in polar co-ordinates (with non-constant basis vectors) is different in that we have to account for the non-constant basis vectors. We have:

$$\frac{\partial \vec{e_r}}{\partial r} = \frac{\partial}{\partial r}\left(\cos\theta.\vec{e_x} + \sin\theta.\vec{e_y}\right)$$
$$= 0 \tag{9.5}$$

But:

$$\frac{\partial \vec{e_r}}{\partial \theta} = \frac{\partial}{\partial \theta}\left(\cos\theta.\vec{e_x} + \sin\theta.\vec{e_y}\right)$$
$$= -\sin\theta.\vec{e_x} + \cos\theta.\vec{e_y} \tag{9.6}$$
$$= \frac{1}{r}\vec{e_\theta}$$

Similarly:

$$\frac{\partial \vec{e_\theta}}{\partial r} = \frac{\partial}{\partial r}\left(-r\sin\theta.\vec{e_x} + r\cos\theta.\vec{e_y}\right)$$
$$= 0.\vec{e_r} + \frac{1}{r}\vec{e_\theta} = \frac{1}{r}\vec{e_\theta} \tag{9.7}$$

And:

$$\frac{\partial \vec{e_\theta}}{\partial \theta} = \frac{\partial}{\partial \theta}\left(-r\sin\theta.\vec{e_x} + r\cos\theta.\vec{e_y}\right)$$
$$= -r\vec{e_r} + 0.\vec{e_\theta} = -r\vec{e_r} \tag{9.8}$$

We notice that the differentials of the basis vectors are a multiple of a basis vector. In general, the differentials of the basis vectors are vectors and are therefore a sum of multiples of the different basis vectors; we have shown this by including the zeros above, (9.7) & (9.8).

The differential of a vector is:

79

$$\frac{\partial \vec{V}}{\partial x^\beta} = \frac{\partial V^\alpha}{\partial x^\beta}\vec{e}_\alpha + V^\alpha \frac{\partial \vec{e}_\alpha}{\partial x^\beta} \tag{9.9}$$

This is just the familiar product rule of differentiation[36].

Christoffel symbols:

We introduce a symbol for the differentials of the basis vectors:

$$\frac{\partial \vec{e}_\alpha}{\partial x^\beta} = \Gamma^\mu_{\alpha\beta}\,\vec{e}_\mu \tag{9.10}$$

Notice that there is summation over the μ. The $\Gamma^\mu_{\alpha\beta}$ is called a Christoffel symbol. It is named after Elwin Bruno Christoffel (1829-1900). A Christoffel symbol is the μth component of the vector that results from differentiating the α basis vector with respect to the β co-ordinate.

The Christoffel symbols, $\Gamma^\mu_{\alpha\beta}$, are a set of $n\times n$ matrices, $\Gamma^\mu_{\sim\beta}$ where n is the dimension of the space; there are n of them corresponding to the values of α. Thus, the Christoffel symbols are linear transformations.

It will transpire that the 4-dimensional space-time covariant derivative is the normal derivative plus a Christoffel symbol or two, and so the 4-dimensional space-time covariant derivative will be a normal derivative with a linear transformation (movement in space) added on. Instead of forming the gauge covariant derivative by adding a bit of the gauge potential, we are forming the 4-dimensional space-time covariant derivative by adding a bit of space-time.

Of course, we might view a potential in gauge space as a bit of that gauge space.

---

[36] We have used the Einstein summation convention.

An affine connection:

An affine connection on a space is usually thought of as defining what is meant by the parallel transport of a vector. In a curved space, the vector will generally change as it is transported from place to place but it will change into another vector; such a vector is a sum of amounts of the basis vectors in the space. The Christoffel symbols are the amounts, the coefficients of the basis vectors, which measure the change during the transport of a vector. If the covariant differential (see below) of a vector is zero as it is transported along a path in the space, we say the vector has been parallel transported.

Within Riemann geometry, it is assumed that the Christoffel symbols are symmetric in their lower indices. This means:

$$\Gamma^{\mu}_{\alpha\beta} = \Gamma^{\mu}_{\beta\alpha} \tag{9.11}$$

Since the theory of GR is based in Riemann geometry, GR also makes this assumption.

Roughly, the Christoffel symbols correspond to the gravitational force.

The 4-dimensional space-time covariant derivative:

Using the Christoffel symbols, the derivative of a vector is:

$$\frac{\partial \vec{V}}{\partial x^{\beta}} = \frac{\partial V^{\alpha}}{\partial x^{\beta}} \vec{e_{\alpha}} + V^{\alpha} \ \Gamma^{\mu}_{\alpha\beta} \ \vec{e_{\mu}} \tag{9.12}$$

We are allowed to replace dummy indices with a different letter giving:

$$\frac{\partial \vec{V}}{\partial x^{\beta}} = \left(\frac{\partial V^{\alpha}}{\partial x^{\beta}} + V^{\mu} \ \Gamma^{\alpha}_{\mu\beta} \right) \vec{e_{\alpha}} \tag{9.13}$$

The components of the vector field are then:

$$D_{\beta}\left(V^{\alpha}\right) = \frac{\partial V^{\alpha}}{\partial x^{\beta}} + V^{\mu} \ \Gamma^{\alpha}_{\mu\beta} \tag{9.14}$$

This is called the 4-dimensional space-time covariant derivative. It is the normal derivative with a bit added.

In some co-ordinate systems, this will reduce to the derivative of a vector which we learned in kindergarten, but it is the proper derivative of a vector in all co-ordinate systems.

Notice that the covariant derivative is not just something that occurs in only curved space. The above polar co-ordinates are over a flat space, but then curved co-ordinates over a flat space are the same as straight co-ordinates in curved space.

Summary:
The 4-dimensional space-time covariant derivative is the normal derivative with a bit of linear transformation added.

Instead of adding a bit of gauge potential, we add a bit of space-time.

The differential in 4-dimensional space-time is just four real numbers division algebra differentials fitted together to make a 4-vector.

Chapter 10

Integration in the Complex Numbers

The 'Fundamental Theorem of Calculus' states that integration is the inverse (reverse) of differentiation. This is certainly the case within the real numbers, but is it the case within the higher-dimensional division algebras? Is the 'Fundamental Theorem of Calculus' fundamental? The answer is yes by definition and also by all that is sensible. Thus, 'The Fundamental Theorem of Calculus' is not a theorem at all; it is simply a statement that says, if we have a function which we can differentiate to form the differential, then that function is the integral of that differential. Integration is the name we will give to the act of moving from the differential to the function we differentiated.

Above, within the complex numbers, we have seen the conventional Cauchy-Riemann differential and we have seen the differential represented by the differential matrix operator. With two types of differentiation, do we have two types of integration, and, in each case, does the 'Fundamental Theorem of Calculus' apply? The answer to this question is also yes. Both types of differentiation can be reversed, and, in both cases, the reversal of differentiation is called integration.

Cauchy-Riemann integration:
There is a whole complicated area of complex analysis dealing with integration within the complex numbers, $\mathbb{C}$, that is founded upon the conventional Cauchy Riemann differential. We do not wish to consider such Cauchy-Riemann integration other than to say that it is the inverse of Cauchy-Riemann differentiation and so the 'Fundamental Theorem of Calculus' applies to Cauchy-Riemann integration. Our concern in this chapter is integration of the matrix differential operator type of differentiation.

Two approaches to integration:

Within all division algebras, including than the real numbers, we have shown there is a differential represented by the differential matrix operator. Considering this type of differential, we can integrate in one of two ways.

a) We can stick with the 'Fundamental Theorem of Calculus' and seek a way of inverting the matrix operator differentiation.

b) We can abandon the 'Fundamental Theorem of Calculus' and consider an integration operator matrix like, within the complex numbers, $\mathbb{C}$:

$$
\begin{bmatrix}
\int_{x_0}^{x_1} \partial x & -\int_{y_0}^{y_1} \partial y \\[2em]
\int_{y_0}^{y_1} \partial y & \int_{x_0}^{x_1} \partial x
\end{bmatrix}
\tag{10.1}
$$

Of course, if we want the inverse of differentiation, we can get that by one means and perhaps also have an integration operator. Perhaps we will have two kinds of integration.

Inverting the differentiation:

The matrix operator differential within the complex numbers, $\mathbb{C}$, is:

$$
\frac{\partial \begin{bmatrix} f(x,y) & g(x,y) \\ -g(x,y) & f(x,y) \end{bmatrix}}{\partial \begin{bmatrix} x & y \\ -y & x \end{bmatrix}}
=
\begin{bmatrix}
\dfrac{\partial f}{\partial x} + \dfrac{\partial g}{\partial y} & \dfrac{\partial g}{\partial x} - \dfrac{\partial f}{\partial y} \\[1.5em]
-\left(\dfrac{\partial g}{\partial x} - \dfrac{\partial f}{\partial y} \right) & \dfrac{\partial f}{\partial x} + \dfrac{\partial g}{\partial y}
\end{bmatrix}
\tag{10.2}
$$

To integrate such a differential, we need to find the two functions $\{f, g\}$ from the differential; for example, suppose we have:

$$\begin{bmatrix} \dfrac{\partial f}{\partial x} + \dfrac{\partial g}{\partial y} & \dfrac{\partial g}{\partial x} - \dfrac{\partial f}{\partial y} \\[2mm] -\left(\dfrac{\partial g}{\partial x} - \dfrac{\partial f}{\partial y}\right) & \dfrac{\partial f}{\partial x} + \dfrac{\partial g}{\partial y} \end{bmatrix} = \begin{bmatrix} 3x^2 y + 4y + x & y - x^3 \\[2mm] -y + x^3 & 3x^2 y + 4y + x \end{bmatrix}$$

$$(10.3)$$

This gives the two equations:

$$\frac{\partial f}{\partial x} + \frac{\partial g}{\partial y} = 3x^2 y + 4y + x \qquad\qquad (1)$$

$$(10.4)$$

$$\frac{\partial g}{\partial x} - \frac{\partial f}{\partial y} = y - x^3 \qquad\qquad (2)$$

By guesswork, we might come to:

$$f = x^3 y \qquad \& \qquad g = 2y^2 + xy + 4 \qquad\qquad (10.5)$$

but there is another way.

Integrating (1) with respect to x and integrating (2) with respect to y gives:

$$f + \int \partial x \frac{\partial g}{\partial y} = x^3 y + 4yx + x^2 + c + m(y) \qquad\qquad (1a)$$

$$\int \partial y \frac{\partial g}{\partial x} - f = \frac{1}{2} y^2 - yx^3 + b + n(x) \qquad\qquad (2a)$$

$$(10.6)$$

Where $\{b, c\}$ are just real numbers and $m(y)$ is a function of only y and $n(x)$ is a function of only x.

Eliminating f by adding $(1a)+(2a)$ gives:

$$\int \partial y \frac{\partial g}{\partial x} + \int \partial x \frac{\partial g}{\partial y} = \left(x^3 y - y x^3\right) + 4yx + x^2 + \frac{1}{2}y^2 + d + m(y) + n(x)$$

$$\int \partial y \frac{\partial g}{\partial x} + \int \partial x \frac{\partial g}{\partial y} = 4yx + x^2 + \frac{1}{2}y^2 + d + m(y) + n(x)$$

$$(10.7)$$

Since we have eliminated the function f on the LHS of the equations, we must have also eliminated the expression of f on the RHS of the equations. We can see in the first equation of (10.7) that:

$$f = x^3 y \qquad (10.8)$$

This concurs with what we know to be the answer, (10.5). Actually, there might be terms like $m(y) + n(x) + const$; we have cheated a little for simplicity of presentation; please forgive your author.

This gives:

$$\frac{\partial f}{\partial x} = 3x^2 y \qquad \& \qquad \frac{\partial f}{\partial y} = x^3 \qquad (10.9)$$

Putting these into (10.4) gives:

$$\frac{\partial g}{\partial x} = y \qquad \& \qquad \frac{\partial g}{\partial y} = 4y + x \qquad (10.10)$$

Integrating (10.10) gives:

$$g(x, y) = xy + s + p(y)$$
$$g(x, y) = 2y^2 + xy + t + q(x) \qquad (10.11)$$

Where $\{s, t\}$ are just real numbers and $p(y)$ is a function of only y and $q(x)$ is a function of only x.

Inspecting (10.11), we see that $p(y) = 2y^2$ and $q(x) = 0$ and $s = t$ giving:

$$g(x, y) = 2y^2 + xy + t \qquad (10.12)$$

This concurs with what we know to be the answer, (10.5). Phew!

The important point is that, because we have two equations and two unknown functions, we can always solve the equations to discover the functions except for a real constant, t. We can invert the differentiation. If we like, we can refer to this inversion of differentiation as integration, and 'The Fundamental Theorem of Calculus' holds. .

Integration by operator:
Consider the complex numbers, $\mathbb{C}$, integration operator:

$$Int = \begin{bmatrix} \int_{x_0}^{x_1} \partial x & -\int_{y_0}^{y_1} \partial y \\ \int_{y_0}^{y_1} \partial y & \int_{x_0}^{x_1} \partial x \end{bmatrix} \tag{10.13}$$

We have included the limits for completeness; we will ignore them in what follows. Note the position of the minus sign. Let this operator act as matrix multiplication upon the differential, (10.3). We will be careful to keep the terms in order; this is not necessary, but it does ease presentation for the reader. We have:

$$\begin{bmatrix} \int_{x_0}^{x_1} \partial x & -\int_{y_0}^{y_1} \partial y \\ \int_{y_0}^{y_1} \partial y & \int_{x_0}^{x_1} \partial x \end{bmatrix} \begin{bmatrix} \dfrac{\partial f}{\partial x} + \dfrac{\partial g}{\partial y} & \dfrac{\partial g}{\partial x} - \dfrac{\partial f}{\partial y} \\ -\left(\dfrac{\partial g}{\partial x} - \dfrac{\partial f}{\partial y}\right) & \dfrac{\partial f}{\partial x} + \dfrac{\partial g}{\partial y} \end{bmatrix}$$

$$= \begin{bmatrix} \int_{x_0}^{x_1} \partial x \left(\dfrac{\partial f}{\partial x} + \dfrac{\partial g}{\partial y}\right) + \int_{y_0}^{y_1} \partial y \left(\dfrac{\partial g}{\partial x} - \dfrac{\partial f}{\partial y}\right) & \sim \\ \int_{y_0}^{y_1} \partial y \left(\dfrac{\partial f}{\partial x} + \dfrac{\partial g}{\partial y}\right) - \int_{x_0}^{x_1} \partial x \left(\dfrac{\partial g}{\partial x} - \dfrac{\partial f}{\partial y}\right) & \sim \end{bmatrix} \tag{10.14}$$

wherein we have squiggled out the duplicate information. Integrating, we have:

$$\begin{bmatrix} f + \int_{x_0}^{x_1} \partial x \, \dfrac{\partial g}{\partial y} + m(y) + c + \int_{y_0}^{y_1} \partial y \, \dfrac{\partial g}{\partial x} - f + n(x) + b & \sim \\[2ex] \int_{y_0}^{y_1} \partial y \, \dfrac{\partial f}{\partial x} + g + l(x) + j - g - \int_{x_0}^{x_1} \partial x \, \dfrac{\partial f}{\partial y} + k(y) + h & \sim \end{bmatrix} \qquad (10.15)$$

We see that the integration operator has not solved the integration but has led to the equation (two off) which we had above, (10.7). We will redo the example above:

$$\begin{bmatrix} \int_{x_0}^{x_1} \partial x & -\int_{y_0}^{y_1} \partial y \\[2ex] \int_{y_0}^{y_1} \partial y & \int_{x_0}^{x_1} \partial x \end{bmatrix} \begin{bmatrix} 3x^2 y + 4y + x & y - x^3 \\[1ex] -y + x^3 & 3x^2 y + 4y + x \end{bmatrix}$$

$$= \begin{bmatrix} x^3 y + 4xy + \dfrac{1}{2} x^2 + m(y) + c - \dfrac{1}{2} y^2 - yx^3 + n(x) + b & \sim \\[2ex] \dfrac{3}{2} x^2 y^2 + 2y^2 + xy + l(x) + j - yx + \dfrac{1}{4} x^4 + k(y) + h & \sim \end{bmatrix}$$

$$(10.16)$$

From this we read off the terms which will disappear:

$$f = x^3 y + m(y) + n(x) + const$$
$$g = xy + l(x) + k(y) + const \qquad (10.17)$$

We differentiate these and compare the result with the known differential, (10.3). We get

$$\frac{\partial f}{\partial x} = 3x^2 y + \frac{\partial n}{\partial x}, \qquad \frac{\partial f}{\partial y} = x^3 + \frac{\partial m}{\partial y}$$

$$\frac{\partial g}{\partial x} = y + \frac{\partial l}{\partial x}, \qquad \frac{\partial g}{\partial y} = x + \frac{\partial k}{\partial y} \qquad (10.18)$$

And:

$$\frac{\partial f}{\partial x}+\frac{\partial g}{\partial y}=3x^2y+\frac{\partial n}{\partial x}+x+\frac{\partial k}{\partial y}=3x^2y+4y+x$$

$$\Rightarrow \frac{\partial n}{\partial x}+\frac{\partial k}{\partial y}=4y$$

(10.19)

Since $n(x)$ is a function of only x, we must have $k(y)=2y^2$ and $n(x)=const$.

And:

$$\frac{\partial g}{\partial x}-\frac{\partial f}{\partial y}=y+\frac{\partial l}{\partial x}-x^3-\frac{\partial m}{\partial y}=y-x^3$$

$$\Rightarrow \frac{\partial l}{\partial x}-\frac{\partial m}{\partial y}=0$$

(10.20)

Since $l(x)$ is a function of only x and $m(y)$ is a function of only y, the implication of (10.20) can be true only if $l(x)=m(y)=const$.

This gives:

$$f = x^3y+const$$
$$g = xy+2y^2+const$$

(10.21)

In matrix form:

$$\begin{bmatrix} x^3y+c & xy+2y^2+d \\ -\left(xy+2y^2+d\right) & x^3y+c \end{bmatrix}$$

(10.22)

We see that the integration operator has solved the equations for us. Above, we proposed two possible types of integration. We see that these two types of integration amount to the same thing and that the 'Fundamental Theorem of Calculus' has endured.

If we wish to impose limits, we simply feed the values into the complex number, (10.22), and we have a complex number.

Looking at the above, we see that integrating the matrix operator differential is much more complicated than integrating a real function.

We will always be faced with solving equations to discover the integral because the matrix operator differential is complicated.

Effectively, matrix operator integration is discovering the potential given the divergence and the curl of that potential.

None-the-less, in matrix operator integration we do not have to worry about contours or residues and the like as we have to when we use Cauchy-Riemann integration. In spite of the calculation difficulties, matrix operator integration is completely comprehensive, and it is conceptually simpler that Cauchy-Riemann integration.

Summary:

Matrix operator integration is the inverse of matrix operator differentiation and so is consistent with 'The Fundamental Theorem of Calculus'.

We have made no reference to contours of integration or to residues or any of the other concepts associated with Cauchy-Riemann integration.

Chapter 11

Integration within the Quaternions

In a previous chapter, we looked at integration within the commutative complex numbers, $\mathbb{C}$. We found that the integration operator is no more than a calculative aid in finding the inverse of the differential. Within the commutative complex numbers division algebra, we found that 'The Fundamental Theorem of Calculus' applies and that integration is the inverse of differentiation. We now seek to integrate within the non-commutative quaternion algebras. We will choose the left-chiral quaternions for our example.

We are seeking a way of doing non-commutative integration. If we can discover non-commutative integration within the quaternions, we can non-commutatively integrate within any non-commutative division algebra.

It will transpire that the integration operator is of no obvious useful application within non-commutative calculus. It will transpire that 'The Fundamental Theorem of Calculus' holds by default. It will transpire that non-commutative integration is a complicated calculation seeking to reverse the non-commutative differentiation.

Non-commutative integration:
There are two ways in which we can approach non-commutative integration.

a) We can begin with both the E-differential and the B-differential and seek a way of calculating the original potential from these two fields. This is to assume that 'The Fundamental Theorem of Calculus' applies and that integration is just inverse differentiation.

b) We can act with a matrix integration operator to produce two integrals. These would be the E-integral and the B-integral.

The second of these, b, is of no obvious meaning, and so our task this chapter is to begin with the E-differential (E-field) and the B-differential (B-field) and to calculate the potential from which these were derived by non-commutative differentiation. This is no easy task.

What is a non-commutative differential?:

We begin by recognising that a non-commutative differential has both a E-differential and a B-differential. If someone gives us only a single field, we immediately know that this field is not a non-commutative differential because a non-commutative differential is two fields.

It might be that one of the expressions is zero; we might be given:

$$E_{[1,1]} = \frac{\partial \phi}{\partial t} + \frac{\partial A_x}{\partial x} + \frac{\partial A_y}{\partial y} + \frac{\partial A_z}{\partial z}$$

$$E_{[1,2]} = \frac{\partial A_x}{\partial t} + \frac{\partial \phi}{\partial x} + \frac{\partial A_z}{\partial y} + \frac{\partial A_y}{\partial z} \qquad \begin{matrix} B_{[1,1]} = 0 \\ B_{[1,2]} = 0 \end{matrix}$$

$$E_{[1,3]} = \frac{\partial A_y}{\partial t} + \frac{\partial A_x}{\partial x} + \frac{\partial \phi}{\partial y} + \frac{\partial A_z}{\partial z} \qquad \begin{matrix} B_{[1,3]} = 0 \\ B_{[1,4]} = 0 \end{matrix} \qquad (11.1)$$

$$E_{[1,4]} = \frac{\partial A_z}{\partial t} + \frac{\partial A_y}{\partial x} + \frac{\partial A_x}{\partial y} + \frac{\partial \phi}{\partial z}$$

We know that this is the result of a commutative differentiation or the differential of a constant potential; this example is actually the differential of a commutative H-type C_4 division algebra.

We will never be given a pair of differential expressions in which the E-differential is zero and the B-differential is non-zero because all division algebra have a real variable and real variables are commutative. All non-commutative division algebras have a centre[37] which is at least one variable.

In the case of two zero fields, the potential is a constant (four constants actually).

Integration as reverse differentiation:
Earlier, we calculated the E-field and the B-field of the potential (5.2) & (5.8) & (5.9):

$$Q_{Pot} = \begin{bmatrix} \phi & -A_x & -A_y & -A_z \\ A_x & \phi & A_z & -A_y \\ A_y & -A_z & \phi & A_x \\ A_z & A_y & -A_x & \phi \end{bmatrix} \tag{11.2}$$

These are:

$$E_{[1,1]} = \frac{\partial \phi}{\partial t} - \frac{\partial A_x}{\partial x} - \frac{\partial A_y}{\partial y} - \frac{\partial A_z}{\partial z}$$

$$E_{[1,2]} = -\frac{\partial \phi}{\partial x} - \frac{\partial A_x}{\partial t}$$

$$E_{[1,3]} = -\frac{\partial \phi}{\partial y} - \frac{\partial A_y}{\partial t} \tag{11.3}$$

$$E_{[1,4]} = -\frac{\partial \phi}{\partial z} - \frac{\partial A_z}{\partial t}$$

---

[37] The centre of an algebra is the set of mutually commutative elements.

$$B_{[1,1]} = 0$$

$$B_{[1,2]} = -\frac{\partial A_y}{\partial z} + \frac{\partial A_z}{\partial y}$$

$$B_{[1,3]} = \frac{\partial A_x}{\partial z} - \frac{\partial A_z}{\partial x} \qquad (11.4)$$

$$B_{[1,4]} = -\frac{\partial A_x}{\partial y} + \frac{\partial A_y}{\partial x}$$

An example:

Suppose we have the left-chiral quaternion potential:

$$POT = \begin{bmatrix} yt + x^2 & xyz & y^2 + xz & z^3 + 4t + 3 \\ -xyz & yt + x^2 & -z^3 - 4t - 3 & y^2 + xz \\ -y^2 - xz & z^3 + 4t + 3 & yt + x^2 & -xyz \\ -z^3 - 4t - 3 & -y^2 - xz & xyz & yt + x^2 \end{bmatrix}$$

$$(11.5)$$

This is:

$$\phi = yt + x^2$$
$$A_x = -xyz$$
$$A_y = -y^2 - xz \qquad (11.6)$$
$$A_z = -z^3 - 4t - 3$$

This potential, (11.5) & (11.6), has E-field and B-field:

$$E = \begin{bmatrix} 3y + yz + 3z^2 & -2x & -t & 4 \\ 2x & 3y + yz + 3z^2 & -4 & -t \\ t & 4 & 3y + yz + 3z^2 & 2x \\ -4 & t & -2x & 3y + yz + 3z^2 \end{bmatrix}$$

$$(11.7)$$

$$B = \begin{bmatrix} 0 & x & -xy & -z+xz \\ -x & 0 & z-xz & -xy \\ xy & -z+xz & 0 & -x \\ z-xz & xy & x & 0 \end{bmatrix} \qquad (11.8)$$

From the E-field and the B-field, we seek expressions for the elements of the potential $\{\phi, -A_x, -A_y, -A_z\}$.

Non-commutative integration:

Non-commutative integration is a complicated computation which begins with the E-differential (E-field) and the B-differential (B-field) and from only these reconstructs the original potential.

Recalling that the E/B differential 1s, (5.1):

$$E = \frac{1}{2}(d_L Q_{Pot} + Q_{Pot} d_R)$$
$$B = \frac{1}{2}(d_L Q_{Pot} - Q_{Pot} d_R) \qquad (11.9)$$

We can get the left-differential and the right-differential by forming the sum and the difference of the E-field and the B-Field; this gives:

$$E + B = d_L Q_{Pot}$$
$$E - B = Q_{Pot} d_R \qquad (11.10)$$

We will need only the left differential. Using (11.10) & (11.7) & (11.8) of this potential, (11.5), d_L to be:

$$d_L = \begin{bmatrix} 3y+yz+3z^2 & -x & -xy-t & xz-z+4 \\ x & 3y+yz+3z^2 & -xz+z-4 & -xy-t \\ xy+t & xz-z4 & 3y+yz+3z^2 & x \\ -xz+z-4 & xy+t & -x & 3y+yz+3z^2 \end{bmatrix}$$
$$(11.11)$$

95

Calculating the spatial potential variables:

We choose to first seek A_x; we use the B-field, (11.4) & (11.8). We will integrate each part of the B-field by the variable with which A_x is differentiated. For example, we will integrate $B_{[1,3]}$ with respect to z and we will integrate $B_{[1,4]}$ with respect to y. Note that there are only two parts of the B-field which have a differential of A_x. We have:

$$\int \partial z \, B_{[1,3]} = A_x - \int \partial z \frac{\partial A_z}{\partial x} + \dots$$

$$= -xyz + j_z(t,x,y) + c_z \tag{11.12}$$

$$\int \partial y \, B_{[1,4]} = -A_x + \int \partial y \frac{\partial A_y}{\partial x} + \dots$$

$$= -yz + xyz + h_y(t,x,z) + c_y \tag{11.13}$$

We have used the expression +... to represent the functions and constants that we have shown explicitly in the second lines of these equations, (11.12) & (11.13).

Adding these two equations, (11.12) & (11.13) will eliminate A_x and the term $-xyz$ and perhaps some parts of the integration functions, $j_z(t,x,y)$ & $h_y(t,x,z)$ and a constant. We therefore know that, being careful of signs:

$$A_x = -xyz + f_x(t,x,y,z) + c \tag{11.14}$$

This concurs with (11.2) & (11.5) above. We now have a partial expression for A_x.

Similarly integrating $B_{[1,2]}$ with respect to z and integrating $B_{[1,4]}$ with respect to x leads to the partial expression:

$$A_y = -xz + f_y(t,x,y,z) + c \tag{11.15}$$

This concurs with (11.2) & (11.5) above.

Similarly integrating $B_{[1,2]}$ with respect to y and integrating $B_{[1,3]}$ with respect to x leads to:

$$A_z = 0 + f_z(t, x, y, z) + c \tag{11.16}$$

This concurs with (11.2) & (11.5) above. We have allowed additional functions of the most general form, $f_i(t, x, y, z)$.

We are part of the way there, but we must be careful.

Feeding the above expressions for $\{A_x, A_y, A_z\}$, (11.14) & (11.15) & (11.16) into the B-field, (11.4), gives:

$$B_{[1,1]} = 0$$

$$B_{[1,2]} = -\frac{\partial A_y}{\partial z} + \frac{\partial A_z}{\partial y} = x + \frac{\partial f_y}{\partial z} + 0 + \frac{\partial f_z}{\partial y}$$

$$B_{[1,3]} = \frac{\partial A_x}{\partial z} - \frac{\partial A_z}{\partial x} = -xy + \frac{\partial f_x}{\partial z} - 0 + \frac{\partial f_z}{\partial x} \tag{11.17}$$

$$B_{[1,4]} = -\frac{\partial A_x}{\partial y} + \frac{\partial A_y}{\partial x} = xz + \frac{\partial f_x}{\partial y} - z + \frac{\partial f_y}{\partial x}$$

Comparing this with the known B-field, (11.8), gives:

$$\frac{\partial f_y}{\partial z} + \frac{\partial f_z}{\partial y} = 0$$

$$\frac{\partial f_x}{\partial z} + \frac{\partial f_z}{\partial x} = 0 \tag{11.18}$$

$$\frac{\partial f_x}{\partial y} + \frac{\partial f_y}{\partial x} = 0$$

We are starting to collate knowledge of the addition integration functions $f_i(t, x, y, z)$.

Using the E-field:
We take the E-field, and we do with it as we did with the B-field. We have:

$$\int \partial t E_{[1,1]} = \phi - \int \partial t \frac{\partial A_x}{\partial x} - \int \partial t \frac{\partial A_y}{\partial y} - \int \partial t \frac{\partial A_z}{\partial z} + \ldots$$

$$= tyz + 3tz^2 + 3ty + f_t(x,y,z) + c$$

$$\int \partial x E_{[1,2]} = -\phi - \int \partial x \frac{\partial A_x}{\partial t} + \ldots = -x^2 + g_x(t,y,z) + c \quad (11.19)$$

$$\int \partial y E_{[1,3]} = -\phi - \int \partial y \frac{\partial A_y}{\partial t} + \ldots = -ty + h_y(t,x,z) + c$$

$$\int \partial z E_{[1,4]} = -\phi - \int \partial z \frac{\partial A_z}{\partial t} + \ldots = 4z + j_z(t,x,y) + c$$

There is nothing conclusive here. All we can do from now on is to guess. We form a trial potential, take the left-differential of this trial potential and compare it to the left-differential of the potential which we calculated above as the sum of the E-field and the B-field, (11.11).

Finding the missing functions:
Using the B-field and the E-field, we have effectively narrowed down the possible potential terms. We now put these terms into a 'trial potential' and calculate the E-field and the B-field of this trial potential. By trial and error, we get our integral. The trial potential is:

$$Trial - Pot =$$

$$\begin{bmatrix} \phi(t,x,y,z) & xyz - f_x & xz - f_y & 0 - f_z \\ -xyz + f_x & \phi(t,x,y,z) & 0 + f_z & xz - f_y \\ -xz + f_y & 0 - f_z & \phi(t,x,y,z) & -xyz + f_x \\ 0 + f_z & -xz + f_y & xyz - f_x & \phi(t,x,y,z) \end{bmatrix} \quad (11.20)$$

Wherein we have omitted the constants which will be lost in differentiation.

The left differential of the trial potential, (11.20), is:

$$d_L (Trial - Pot)_{[1,1]} = \frac{\partial \phi}{\partial t} + yz - \frac{\partial f_x}{\partial x} - \frac{\partial f_y}{\partial y} - \frac{\partial f_z}{\partial z}$$

$$d_L (Trial - Pot)_{[1,2]} = -\frac{\partial f_x}{\partial t} - \frac{\partial \phi}{\partial x} - \frac{\partial f_z}{\partial y} + x - \frac{\partial f_y}{\partial z}$$

$$d_L (Trial - Pot)_{[1,3]} = -\frac{\partial f_y}{\partial t} - \frac{\partial f_z}{\partial x} - \frac{\partial \phi}{\partial y} - xy - \frac{\partial f_x}{\partial z}$$

$$d_L (Trial - Pot)_{[1,4]} = -\frac{\partial f_z}{\partial t} - z - \frac{\partial f_y}{\partial x} + xz - \frac{\partial f_x}{\partial y} - \frac{\partial \phi}{\partial z}$$

$$(11.21)$$

We can simplify this a little by using the result (11.18):

$$d_L (Trial - Pot)_{[1,1]} = \frac{\partial \phi}{\partial t} + yz - \frac{\partial f_x}{\partial x} - \frac{\partial f_y}{\partial y} - \frac{\partial f_z}{\partial z}$$

$$d_L (Trial - Pot)_{[1,2]} = -\frac{\partial f_x}{\partial t} - \frac{\partial \phi}{\partial x} + x$$

$$d_L (Trial - Pot)_{[1,3]} = -\frac{\partial f_y}{\partial t} - \frac{\partial \phi}{\partial y} - xy$$

$$d_L (Trial - Pot)_{[1,4]} = -\frac{\partial f_z}{\partial t} - z + xz - \frac{\partial \phi}{\partial z}$$

$$(11.22)$$

Comparing these results, (11.22), with the left-differential we calculated as the sum of the E-field and the B-field, (11.11), we are driven to:

$$\frac{\partial \phi}{\partial t} - \frac{\partial f_x}{\partial x} - \frac{\partial f_y}{\partial y} - \frac{\partial f_z}{\partial z} = 3z^2 + 3y$$

$$-\frac{\partial f_x}{\partial t} - \frac{\partial \phi}{\partial x} = -2x$$

$$-\frac{\partial f_y}{\partial t} - \frac{\partial \phi}{\partial y} = -t$$

$$-\frac{\partial f_z}{\partial t} - \frac{\partial \phi}{\partial z} = 4$$

$$(11.23)$$

The reader might check that this is consistent with what we know to be the answer, (11.6).

As is often the case when calculating integrals, we are now down to guesswork. Since we know the answer, guessing is easy. If we did not know the answer, guessing would be very hard.

Summary of non-commutative integration:
We have assumed 'The Fundamental Theorem of Calculus' that integration is the reverse of differentiation. Even so, we are unable to calculate the non-commutative integral without guesswork.

The important point is:
We have rejected any idea that the non-commutative integral would produce two integrals, an E-integral and a B-integral. Non-commutative integration is no more than reversal of non-commutative differentiation. We begin with an E-field and a B-field, and we seek to calculate the potential of which these are the E-field and the B-field.

Chapter 12

Concluding Remarks

We have discovered how to do calculus in every type of space. We have discovered how to do calculus in the division algebra spaces, both commutative and non-commutative. We have discovered how to do calculus in the 4-dimensional space-time of our universe. We have covered how to do gauge covariant differentiation within the fibre bundle that is the spatial structure of our universe.

There are no more kinds of space, and so there is no more calculus.

In all cases, we have taken integration to be the reverse (inverse) of differentiation. The view that integration is the inverse (reverse) of differentiation is called the 'Fundamental Theorem of Calculus'.

We began by realising that the differential is a product. From this it follows that the differential exists in only division algebras. We discovered how to differentiate commutatively within every commutative division algebra. We discovered how to differentiate non-commutatively within every non-commutative division algebra.

We then discovered how to sum the differentials within isomorphic division algebras to form the superimposed differential and how to adjust the superimposed differential by adding a bit of gauge potential to it to form the gauge covariant derivative. Thus, we discovered how to do calculus in the fibre bundle which is the spatial structure of our universe.

We then discovered that differentiation in our 4-dimensional space-time is just a vector sum of four differentials within the real numbers together with an adjustment to account for the curvature of our 4-dimensional space-time.

We would like to say "Job done", but there are many details still to be clarified and there is much new ground to be explored. Perhaps the reader will play a part in bringing such clarification and exploration to humankind.

Along the way, we found a kind of differentiation within the complex numbers, $\mathbb{C}$, which is manifestly more comprehensive, simpler, and obviously the correct form of differentiation with a division algebra. Thus, we discarded the Cauchy-Riemann calculus that has been used within the complex numbers for some 200 years or so[38].

We have a very high opinion of ourselves as mathematicians. We take mathematics to be solidly built upon very secure foundations; we take it to be unshakably correct. The realisation of the incorrect nature of the Cauchy-Riemann calculus shows how easily we can make errors. For 200 years, mathematicians both brilliant and mundane walked down the wrong path cheerfully confident in their godlike understanding of mathematics. We saw ourselves as the keepers of absolute truth. Clearly, even such gods as we can make errors. Let us be humbled.

There is a view throughout modern mathematics and modern physics that we 'are in a period of consolidation'. The outstanding success of quantum electro-dynamics and the discovery of the Higgs boson at CERN are taken to confirm that we at last have the correct understanding of the universe and that all that remains is to sort out a few details. The recent discovery of gravitational waves adds to this feeling of this being a 'consolidation period'. Yet there is much which we do not understand. Your author opines that within the next decade or two, our arrogant confidence in the 'consolidation period' will be shaken to the core. There are many seeds of collapse growing within our understanding of physics and of mathematics.

The history of human endeavour is replete with instances of humankind taking the wrong path only to later find ourselves tangled in brambles or waist deep in marshland. We do it in science; we do it in economics; we do it in politics; we do it in industry; we do it in personal relationships. Well! it seems we also do it in mathematics. We are as blind people stumbling and groping our way forward along an unlit country lane on a moonless midnight. We are buoyed up by only our

---

[38] Such a huge matter dismissed in such a small paragraph!

arrogance and self-belief. Yet, if we turn our heads and look back along the way we have walked, other than for the need of a few steps backward, the road we have travelled is well lit and well metalled.

We have come a long way from the 'first mathematician'.

I hope this book has clarified calculus for the reader, and I hope I have been able to present the material clearly and simply. It has been a pleasure writing for such kind and gentle folks as your good-selves.

Dennis Morris

Brotton

July 2016

Bilbliography

Bak, J. & Newman, Donald, J. : Complex Analysis

Lang, Serge : A First Course in Calculus

MacMahon, David : Quantum Field Theory DeMystified

Ward, J. P. : Quaternions and Cayley Numbers

Other Books by the Same Author

The Naked Spinor – a Rewrite of Clifford Algebra

Spinors exist in Clifford algebras. In this book, we explore the nature of spinors. This book is an excellent introduction to Clifford algebra.

Complex Numbers The Higher Dimensional Forms – Spinor Algebra

In this book, we explore the higher dimensional forms of complex numbers. These higher dimensional forms are closely connected to spinors.

Upon General Relativity

In this book, we see how 4-dimensional space-time, gravity, and electromagnetism emerge from the spinor algebras. This is an excellent and easy paced introduction to general relativity.

From Where Comes the Universe

This is a guide for the lay person to the physics of empty space.

Empty Space is Amazing Stuff – The Special Theory of Relativity

This book deduces the theory of special relativity from the finite groups. It gives a unique insight into the nature of the 2-dimensional space-time of special relativity.

The Nuts and Bolts of Quantum Mechanics

This is a gentle introduction to quantum mechanics for undergraduates.

Quaternions

This book pulls together the often separate properties of the quaternions. Non-commutative differentiation is covered as is non-commutative rotation and non-commutative inner products along with the quaternion trigonometric functions.

The Uniqueness of our Space-time

This book reports the finding that the only two geometric spaces within the finite groups are the two spaces which together form our universe. This is a startling finding. The nature of geometric space is explained alongside the nature of division algebra space. This book is a catalogue of the higher dimensional complex numbers up to dimension fifteen.

Lie Groups and Lie Algebras

This book presents Lie theory from a diametrically different perspective to the usual presentation. This makes the subject much more intuitively obvious and easier to learn. Included is perhaps the clearest and simplest presentation of the true nature of the Lie group $SU(2)$ ever presented.

The Physics of Empty Space

This book presents a comprehensive understanding of empty space. The presence of 2-dimensional rotations in our 4-dimensional space-time is explained. Also included is a very gentle introduction to non-commutative differentiation. Classical electromagetism is deduced from the quaternions.

The Electron

This book presents the quantum field theory view of the electron and the neutrino. This view is radically different from the classical view of the electron presented in most schools and colleges. This book gives a very clear exposition of the Dirac equation including the quaternion rewrite of the Dirac equation. This is an excellent introduction to particle physics for students prior to university, during university and after university courses in physics.

The Quaternion Dirac Equation

This short book derives a quaternion form of the Dirac equation. The convention Dirac equation is set in a 16-dimensional Clifford algebra. The 4-dimensional quaternion form of the Dirac equation is much simpler and mathematically more straight-forward than the conventional Dirac equation. The quaternion Dirac equation leads to a non-chiral (spin either up or down) massive electron field and to a massless neutrino field. The massless neutrino field is chiral leading to only left-chiral neutrinos. Although the neutrino field is massless, allowing neutrinos to travel at the speed of light, the neutrino field squared is massive allowing neutrino oscillation.

An Essay on the Nature of Space-time

This small and inexpensive volume presents a view of the nature of empty space using only simple mathematics. The expansion of the universe is explained, and a, somewhat mathematical, explanation of dark energy is presented.

Even Mathematicians and Physicists make Mistakes

This book points out what seems to be several important errors of modern physics and modern mathematics. Errors like the misunderstanding of rotation, the failure to teach the higher dimensional complex numbers in most universities, and the mathematical inconsistency of the Dirac equation and some casual errors are discussed. These errors are set in their historical circumstances and there is discussion about why they happened and the consequences of their happening. There is also an interesting chapter on the nature of mathematical proof within our society, and several famous proofs are discussed (without the details).

Finite Groups – A Simple Introduction

This book introduces the reader to finite group theory. Many introductory books on finite groups bury the reader in geometrical examples or in other types of groups and lose the central nature of a finite group. This book sticks firmly with the permutation nature of finite groups and elucidates that nature by the extensive use of permutation matrices. Permutation matrices simplify the subject considerably. This book is probably unique in its use of permutation matrices and therefore unique in its simplicity.

The Left-handed Spinor

This book covers the left-handed parts of mathematics which we call the chiral algebras. These algebras have CP invariance, violation of parity, and many other aspects which makes them relevant to theoretical physics. It is quite a revelation to discover that mathematics is left-handed.

Non-commutative Differentiation and the Commutator

(The Search for the Fermion Content of the Universe)

This book develops the theory of non-commutative differentiation from the fundamentals of algebra. We see what an algebraic operation (addition, multiplication) really is, and we discover that the commutator is a third fundamental algebraic operation within some division algebras. This leads to the first part of the derivation of the fermion content of the universe.

Index

W

weak force, 44, 48
weak force tensor, 56

Z

zero divisors, 21